MEDICAL DEVICE RELIABILITY AND ASSOCIATED AREAS

MEDICAL DEVICE RELIABILITY AND ASSOCIATED AREAS

B.S. Dhillon, Ph.D.

CRC Press
Boca Raton London New York Washington, D.C.

Library of Congress Cataloging-in-Publication Data

Catalog record is available from the Library of Congress.

Direct all inquiries to CRC Press LLC, 2000 N.W. Corporate Blvd., Boca Raton, Florida 33431.

International Standard Book Number 0-8493-0312-5
Printed in the United States of America 1 2 3 4 5 6 7 8 9 0
Printed on acid-free paper

Preface

The history of modern medical devices and equipment may be traced back to as early as the 17th century. For example, with respect to the invention of the thermometer, in 1603 Galileo developed a device to measure temperature which was later improved by Sanatoria Santonio so that it can measure the temperature of the human body. In 1993, worldwide sales for the U.S. medical device industry reached approximately $41 billion — 5.8% more than 1992 sales of $38.8 billion. This growth makes the medical device industry one of the fastest growing activities in the U.S. economy.

Although the history of reliability engineering can be traced back to World War II, the application of reliability engineering concepts to medical devices is a fairly recent idea that goes back to the latter part of the 1960s when many publications on medical device reliability emerged.

Today, a large number of books on general reliability have been published. Only a few discuss medical device reliability but they lack a comprehensive and current coverage of developments in the field. The main objective of this book is to cover current developments in medical device reliability and associated areas, and provide useful tools and techniques to professionals working in the medical device industry. This book fulfills a vital need and puts emphasis on the structure of concepts rather than mathematical rigor and minute details. It includes a comprehensive list of references on medical device reliablity and associated areas, contains examples and solutions, and at the end of each chapter, problems to test the reader's comprehension. The terms "medical device" and "medical equipment" are interchangeably used in the book.

Chapter 1 presents various introductory aspects of medical device reliability and related areas including historical developments, the need for reliability in medical devices, medical device related facts and figures, and government control and liability.

Chapter 2 is devoted to basic mathematics and concepts that are useful for performing reliability and associated analysis of medical devices.

Chapter 3 presents tools for assuring medical device reliability such as failure mode and effects analysis (FMEA), fault tree analysis (FTA), failure rate evaluation and parts count methods, Markov method, and common-cause failure analysis method.

Chapter 4 discusses human error in health care systems and covers topics such as human error related facts and figures in health care, causes of patient injuries, medical device related operator errors, and useful human error analysis methods.

Chapter 5 presents various aspects of medical device software quality assurance and reliability. The coverage ranges from software terms and definitions to mathematical models for predicting reliability of medical device software.

The topic of medical device safety assurance is presented in Chapter 6. The chapter discusses the types of medical device safety, essential safety requirements for medical devices, safety in device life cycle, and safety analysis methods.

Chapters 7 and 8 are devoted to medical device risk assessment and control, and quality assurance, respectively. Under risk assessment and control, some of the topics covered are risk analysis process, risk analysis methods, advantages of risk analysis, and case studies of risk assessment in medical devices. Similarly, quality assurance includes topics such as regulatory compliance of medical device quality assurance, medical device design quality assurance program, total quality management, and tools for assuring medical device quality.

Chapter 9 presents various areas concerning medical device reliability testing including accelerated testing, success testing, and meantime between failures (MTBF) calculation methods.

The subject of medical device costing is discussed in Chapter 10. Some of the topics covered in the chapter include reasons for device costing, investment decision making methods, reliability related cost estimation models, and life cycle costing.

Chapter 11 presents two main topics concerning medical devices: maintenance and maintainability. It also covers topics such as indices for repair and maintenance, ventilator maintenance, models for medical equipment maintenance, maintainability design factors, and maintainability measures.

Chapter 12 covers three important areas concerning medical devices: reliability related standards, failure data sources, and failure data and analysis. In particular, the chapter contains data on various aspects concerning medical devices and Bartlett and maximum likelihood estimation methods.

This book is useful to biomedical engineers, design managers concerned with the development of biomedical equipment and devices, biomedical engineering senior level undergraduate and graduate students, biomedical equipment users, college and university teachers, and short biomedical equipment and device reliability course instructors and students.

B.S. Dhillon
Ottawa, Ontario

About the Author

Dr. B.S. Dhillon is professor of mechanical engineering at the University of Ottawa where he has been teaching for more than 20 years. He has served as a chairman and director of the mechanical engineering department/engineering management program for more than 10 years at the University of Ottawa. Dr. Dhillon has published more than 270 articles on reliability and maintainability engineering and related subjects. He is on the editorial boards of five international journals on reliability. In addition, Dr. Dhillon has written 21 books on various aspects of system reliability, safety, human factors, design, and engineering management published by Wiley (1981), Van Nostrand (1982), Butterworth (1983), Marcel Dekker (1984), and Pergamon (1986). His books on reliability have been translated into several languages including Russian, Chinese, and German. He has served as general chairman of two international conferences on reliability and quality control held in Los Angeles and Paris in 1987.

Dr. Dhillon has served as a consultant to the Ottawa Heart Institute in the design and development of medical devices for many years. He attended the University of Wales where he received a B.S. degree in electrical and electronics engineering and an M.S. degree in mechanical engineering. He also received a Ph.D. in industrial engineering from the University of Windsor.

Dr. Dhillon has received the American Society of Quality Control Austin J. Bonis Reliability Award, the Society of Reliability Engineers' Merit Award, the Gold Medal of Honour (American Biographical Institute), and Faculty of Engineering Glinski Award for Excellence in Reliability Engineering Research. He is a registered professional engineer in Ontario and is listed in the *American Men and Women of Science*, *Men of Achievements, International Dictionary of Biography*, *Who's Who in International Intellectuals*, and *Who's Who in Technology.*

Table of Contents

Chapter 11

Acknowledgments

I am indebted to my friends, colleagues, and students for their interest and encouragement throughout this project. In particular, I am grateful to my dear friend Dr. Tofy Mussivand, President of the World Heart Corporation for introducing me to the reliability problems of modern medical devices. I also thank my children, Jasmine and Mark, for their patience. Last, but not least, I thank my wife, Rosy, for her editorial input, patience, tolerance and her coffee-making!

Dedication

This book is affectionately dedicated to the memory of my late grandfather Dulip S. Mann who arrived in Canada 100 years ago.

1 Introduction

CONTENTS

1.1 HISTORICAL DEVELOPMENTS

The history of the use of medical devices may be traced back to the ancient times. For example when the ancient Egyptians and Etruscans used dental devices.[1]

Today, in many countries, health care expenditures are among the largest social costs, and over the past three decades they have increased quite rapidly. In 1980, Organization for Economic Cooperation and Development (OECD) countries on average spent 4.2% of their gross domestic product (GDP) on health care, and by 1984 the figure increased to almost 8% of GDP.[2]

In 1958, sales of medical devices in the U.S. totaled less than $1 billion, and grew to more than $17 billion in 1983.[3,4] Furthermore, in 1988, the U.S. medical equipment production reached around $22 billion.[2] It means the production of medical devices/equipment is an important sector in the U.S. industry.

The history of the reliability field may be traced back to the 1930s and 1940s, when the probability concepts were applied to electric power generation related problems[5-8] and Germans applied the basic reliability concepts to improve reliability of their V1 and V2 rockets.[9,10] Ever since those days, many new developments have taken place, and the field has branched out into many specialized areas: software reliability, human reliability, mechanical reliability, power system reliability , structural reliability, etc. Comprehensive lists of publications on almost all of the reliability areas are given in references 11 and 12.

The real beginning of the medical device reliability field may be regarded as the latter part of the 1960s. During this period, several publications on the subject appeared.[13-17] An article[18] published in 1980 listed most of the publications on the subject, and in 1983 a text on reliability devoted a chapter to medical device/equipment reliability.[9] Nowadays, the medical device reliability field has become an important component of the general field.

This chapter presents some introductory aspects of the medical device reliability field.

1.2 NEED OF RELIABILITY IN MEDICAL DEVICES

A modern hospital uses more than 5000 types of medical devices, ranging from a simple tongue depressor to a complex pacemaker. The criticality of their reliability may vary from device to device. The failure of medical devices in the past due to hardware and other reasons has been very costly in terms of fatalities, injuries, dollars and cents, etc. Furthermore, some of today's medical devices have become very complex and are expected to operate under a stringent environment, thus putting more pressure on their reliability assurance during the design phase. Other instrumental factors requiring better reliability of medical devices include:

- Liability suits
- Public pressure
- Government regulations: one typical example of such regulations is the Medical Device Amendments to the Federal Food, Drug and Cosmetic Act in 1976, thus empowering the Food and Drug Administration (FDA) to regulate medical devices during their design and development phases. Since 1976, the FDA demands that all medical devices be safe and effective for their intended purpose and in fact in the mid 1980s, the FDA added the term reliability to its original demand. Subjecting a medical device to a reliability program provides a systematic approach to the product development process and assures that the regulatory requirements are adequately satisfied. It also gives confidence to a certain degree that inspection by regulatory bodies will not lead to major discrepancies. All in all, improved reliability medical devices will be safe, cost-effective, and easy to maintain.

1.3 MEDICAL DEVICE-RELATED FACTS AND FIGURES

Some of the facts and figures directly or indirectly associated with medical devices are as follows.

- In 1983, there were at least 3000 companies involved with medical devices in the U.S., and the medical device industry employed approximately 200,000 people.[3,4]
- Currently the American Hospital Association (AHA) has approximately 5000 institutional, 600 associate, and 40,000 personal members.[19] The institutional members include hospitals, headquarters of the health care

systems, hospital-affiliated educational programs, etc. The associate members include commercial companies, suppliers, and consultants. The personal members include individuals working in the health care field, health care executive assistants, and students.

- In 1997, there were 10,420 registered manufacturers involved in the manufacture of medical devices in the U.S.[20]
- In 1974, the total assets of the AHA-registered hospitals were valued at around $52 billion.[21]
- A study conducted around the middle of 1990s discovered that 93% of medical devices in the U.S. had markets worth below $150 million.[22] During the early 1980s, the U.S. exported more than $2 billion worth of medical devices annually.[4,23]
- In 1969, the U.S. Department of Health, Education, and Welfare special committee reported that over ten years, 10,000 injuries were associated with medical devices and 731 resulted in death.[4,24]
- In 1988, the world medical equipment production was valued at $36.1 billion,[1] and in 1997, the world market for medical devices was estimated to be about $120 billion.[25]
- Faulty instrumentation accounts for 1200 deaths per year in the U.S.[26,27]
- More than 50% of all technical medical equipment problems are due to operator errors.[28]
- The top ten orthopedic companies in the U.S. increased their regulatory affairs staff by 39% during the early years of the 1990s.[29]
- In 1990, a study conducted by the FDA reported that approximately 44% of the quality-related problems that led to the voluntary recall of medical devices from October 1983 to September 1989 were due to deficiencies/errors that could have been prevented through effective design controls.[30]
- Emergency Care Research Institute (ECRI) tested a sample of 15,000 various hospital products and found that 4% to 6% of them were sufficiently dangerous to warrant immediate correction.[28]
- In 1990, the U.S. Congress passed the Safe Medical Device Act (SMDA), which helped strengthen the FDA to implement the Preproduction Quality Assurance Program. This program encourages and requires manufacturers of medical devices to address deficiencies in product design that contribute to malfunctions.

1.4 GOVERNMENT CONTROL AND LIABILITY

As medical devices have become complex and widely used and their use more critical, many governments around the world regulate their design, manufacture, use, and import, including quality and reliability. In the U.S., the FDA is empowered to regulate medical devices to ensure their safety and effectiveness.[30]

The history of the U.S. government's direct or indirect involvement with medical devices may be traced back to the passage of the Biologics Act of 1902 and the Food and Drugs Act of 1906.[1] The Federal Food, Drug, and Cosmetic Act of 1938 authorized the FDA to take formal or informal regulatory measures against the misbranding or adulteration of medical devices. Misbranding included factors such as false or misleading labeling, failure of the label to contain the name and address of the manufacturer, and unsatisfactory directions for use and warnings for misuse. Adulteration means the device contained filthy material, was prepared under unsatisfactory sanitary condition, or differed from the quality claimed on the label.

In 1968, Congress passed the Radiation Control for Health and Safety Act to control harmful radiation emissions from electronic products, and in 1971 delegated responsibility for its administration to the FDA.[1] Amendments to the Federal Food, Drug and Cosmetic Act in 1976 empowered the FDA to regulate medical devices during their design and development phases.

In 1984, Congress passed the Cardiac Pacemaker Registry Act to assist the Health Care Financing Administration (HCFA) in determining proper Medicare payments as well as to assist FDA in determining the compliance of the devices with regulatory requirements. The Safe Medical Device Act (SMDA) was passed in 1990 and authorized the FDA to implement the Preproduction Quality Assurance Program. More specifically, the program requires and encourages device manufacturers to address deficiencies that lead to failure during design.

Nowadays, similar types of regulations concerning medical devices are also being followed in other countries. For example, the Medical Device Directive (MDD) of the European Union (EU) outlines the requirements for medical devices in the EU countries.[31]

Product liability is also an important factor for manufacturing safe and reliable medical devices, particularly in the U.S. For example, the U.S. Supreme Court's ruling on Medtronic, Inc. vs. Lohr, in 1996 has put tremendous pressure on manufacturers to produce safe and reliable devices. In the ruling, the court rejected Medtronic's claim that federal law gives immunity to manufacturers from all civil liability, including when people are killed or injured due to their defective products.[32]

This case was filed by Lora Lohr of Florida who was implanted with a cardiac pacemaker manufactured by Medtronic to correct her irregular heart rhythm. The pacemaker failed and resulted in emergency surgery, as well as additional procedures. Thus, in the lawsuit, Lohr alleged that the pacemaker failure was due to defective design, manufacturing, and labeling.

Court cases such as this have served as a catalyst for manufacturers to produce safe and reliable medical devices.

1.5 MEDICAL DEVICE RECALLS

Defective medical devices are subject to recall in the U.S., and the *FDA Enforcement Report* regularly publishes data on such recalls. For example, a total of 230 medical device-related recalls were made from 1980 to 1982.[28] They were categorized into nine problems areas: defects in material selection and manufacturing (54), faulty product design (40), contamination (40), mislabeling (27), radiation (X-ray) violations (25), electrical problems (17), defective components (13), misassembly of parts (10), and no premarket approval and failure to comply with good manufacturing practices (GMPs) (4).

The first four categories, i.e., defects in material selection and manufacturing, faulty product design, contamination, and mislabeling, accounted for 70% of the recalls. The components of the defects in material selection and manufacturing classification included manufacturing defects, inappropriate materials, and material deterioration. The subcategories of the faulty product design classification were alarm defects, premature failure, electrical interference, potential for leakage of fluids into electrical components, potential for malfunction, etc. The elements of the contamination category included defective package seals, other package defects and nonsterility. There were four subcategories of the mislabeling category: incomplete labeling, misleading or inaccurate labeling, inadequate labeling, and disparity between label and product.

Similar information on other medical device recalls is available from the FDA.

1.6 MEDICAL EQUIPMENT CLASSIFICATION

As the health care system uses a large variety of electronic equipment, it can be grouped into three major classifications:[17] A, B, and C.

Classification A includes medical equipment/devices that are directly and immediately responsible for the patient's life or may become so in emergencies. More specifically, when this type of equipment fails, there is seldom sufficient time for repair. Therefore, this equipment must always operate at the moment of need. It must have high reliability. Some examples of the Classification A equipment are cardiac pacemakers, cardiac defibrillators, respirators, and electro-cardiographic monitors.

Classification B contains a vast majority of medical equipment used for routine or semi-emergency diagnostic or therapeutic purposes. Failure of such equipment does not result in the same emergency as in the case of Classification A equipment because there is time for repair. Some of the devices that fall under this classification are ultrasound equipment, spectrophotometers, electrocardiograph and electroencephalograph recorders and monitors, gas analyzers, colorimeters, and diathermy equipment.

Classification C contains equipment that is not critical to a patient's life or welfare but simply serves as a convenience equipment. Two examples of the Classification C equipment are wheelchairs and bedside television sets.

All in all, there could be some overlap between these three classifications of equipment, particularly between Classifications A and B.

After the passage of the Medical Device Amendments of 1976, the FDA classified devices marketed prior to 1976 into the following three categories:[4]

- **Category I**. This contained devices in which general controls such as good manufacturing practices were considered satisfactory with respect to safety and efficacy.
- **Category II**. This contained devices in which general controls were considered insufficient with respect to safety and efficacy and in which performance standards could be established.

- **Category III**. This contained devices in which the manufacturer is required to submit evidence of safety and efficacy with the aid of well-designed studies. More specifically, the devices included in this category support life, prevent health impairment, or present an unreasonable risk of injury or illness and need FDA approval prior to their marketing.

1.7 MEDICAL DEVICE RELIABILITY AND ASSOCIATED AREAS

To improve the reliability of medical devices, one has to consider not only their reliability aspect but also their associated areas. These areas include quality, safety, risk assessment, maintenance, maintainability, and cost. Areas such as these directly or indirectly influence the reliability of medical devices. For example, a poorly maintained medical device can significantly influence its reliability.

It may be said that in order to study the reliability of medical devices, its associated areas must also be understood. Therefore, some portions of this book are devoted to these areas.

1.8 TERMS AND DEFINITIONS

In reliability work, many terms and definitions are used. This section presents selective terms and definitions directly or indirectly concerned with the medical device reliability and its associated areas. Their understanding is considered essential to comprehending the material presented in this book. The terms and definitions presented below were taken from various sources.[9,33-48]

- **Medical device**. This is any instrument, machine, implant, *in vitro* reagent, apparatus, contrivance, implement, or other related or similar article, including any part, component, or accessory that is intended for application in diagnosing diseases or other conditions or in the cure, treatment, mitigation, or prevention of disease or intended to affect the structure of any function of the body.[33]
- **Reliability**. This is the probability that an item will carry out its function satisfactorily for the stated period when used according to the specified conditions.
- **Maintainability**. This is the probability that a failed item will be repaired or restored to its satisfactory operational condition.
- **Risk**. This is the chance, degree of probability, or possibility of loss, damage, harm, or injury.
- **Maintenance**. This is all measures appropriate for retaining an item in, or repairing/restoring it to, a stated state, ensuring that physical assets continue to meet their stated missions.
- **Quality**. There are many definitions of quality, and one of them is simply conformance to requirements.[36]

- **Safety**. This is conservation of human life and its effectiveness, and the prevention of damage to items as per operational requirements.
- **Cost**. This is the money paid or payable for the acquisition of property, materials, or services.
- **Failure**. This is the inability of an item to operate within the stated guidelines.
- **Human error**. This is the failure to carry out a required task (or the performance of a prohibited action) that could result in disruption of scheduled operations or damage to equipment or property.
- **Mean time to failure**. This is in the case of exponentially distributed times to failures, the sum of the operating time of given items over the total number of malfunctions or failures.
- **Mission time**. This is the time during which the item is carrying out its stated mission.

1.9 PROBLEMS

1. Discuss the need of reliability in medical devices.
2. Write an essay on facts and figures relating to medical devices.
3. Discuss the actions relating to government control in the U.S. with respect to medical devices.
4. Describe the issue of liability with respect to medical devices.
5. What are the basic reasons for the medical device recalls in the U.S.?
6. Discuss classifications of the electronic medical equipment.
7. Discuss the following items concerning medical devices:
    - Risk assessment
    - Safety
8. Discuss medical device reliability vs. medical device quality.
9. Define the following terms:
    - Medical device
    - Failure
    - Reliability
10. Write an essay on the history of medical device/equipment reliability.

REFERENCES

1. Hutt, P.B., A History of Government Regulation of Adulteration and Misbranding of Medical Devices, *The Medical Device Industry*, N.F. Estrin, Ed., Marcel Dekker Inc., New York, 1990, pp. 17–33.
2. Fuchs, M.C., Economics of U.S. Trade in Medical Technology and Export Promotion Activities of the U.S. Department of Commerce, *Medical Device Technology*, N.F. Estrin, Ed., Marcel Dekker Inc., New York, 1990, pp. 917–928.
3. Federal Policies and the Medical Devices Industry, Office of Technology Assessment, U.S. Government Printing Office, Washington, DC, 1984.

4. Banta, H.D., The Regulation of Medical Devices, *Preventive Medicine*, Vol. 19, 1990, pp. 693–699.
5. Lyman, W.J., Fundamental Consideration in Preparing a Master System Plan, *Electrical World*, Vol. 101, 1933, pp. 778–792.
6. Smith, S.A., Service Reliability Measured by Probabilities of Outage, *Electrical World*, Vol. 103, 1934, pp. 371–374.
7. Dhillon, B.S., *Power System Reliability, Safety and Management*, Ann Arbor Science Publishers, Ann Arbor, MI, 1983.
8. Coppola, A., Reliability Engineering of Electronic Equipment: A Historical Perspective, *IEEE Transactions on Reliability*, Vol. 33, 1984, pp. 29–35.
9. Dhillon, B.S., *Reliability Engineering in Systems Design and Operation*, Van Nostrand Reinhold Company, New York, 1983.
10. Shooman, M.L., *Probabilistic Reliability: An Engineering Approach*, McGraw-Hill Book Company, New York, 1968.
11. Dhillon, B.S., *Reliability Engineering Applications: Bibliography on Important Applications Areas*, Beta Publishers, Gloucester, Ontario, 1992.
12. Dhillon, B.S., *Reliability and Quality Control: Bibliography on General and Specialized Areas*, Beta Publishers, Gloucester, Ontario, 1992.
13. Johnson, J.P., Reliability of ECG Instrumentation in a Hospital, *Proceedings of the Annual Symposium on Reliability*, 1967, pp. 314–318.
14. Gechman, R., Tiny Flaws in Medical Design Can Kill, *Hosp. Top.*, Vol. 46, 1968, pp. 23–24.
15. Meyer, J.L., Some Instrument Induced Errors in the Electrocardiogram, *J. Am. Med. Assoc.*, Vol. 201, 1967, pp. 351–358.
16. Taylor, E.F., The Effect of Medical Test Instrument Reliability on Patient Risks, *Proceedings of the Annual Symposium on Reliability*, 1969, pp. 328–330.
17. Crump, J.F., Safety and Reliability in Medical Electronics, *Proceedings of the Annual Symposium on Reliability*, 1969, pp. 320–322.
18. Dhillon, B.S., Bibliography of Literature on Medical Equipment Reliability, *Microelectronics and Reliability*, Vol. 20, 1980, pp. 737–742.
19. Dhillon, B.S., *Design Reliability*, CRC Press, Boca Raton, FL, 1999.
20. Allen, D., California Home to Almost One-Fifth of U.S. Medical Device Industry, *Med. Dev. Diag. Ind. Mag.*, Vol. 19, No. 10, 1997, pp. 64–67.
21. Fairhurst, G.F., Murphy, K.L., Help Wanted, *Proceedings of the Annual Reliability and Maintainability Symposium*, 1976, pp. 103–106.
22. Bethune, J., The Cost Effectiveness Bugaboo, *Med. Dev. Diag. Ind. Mag.*, Vol. 19, No. 4, 1997, pp. 12–15.
23. Highlights of the U.S. Export and Import Trade, Bureau of Census, U.S. Department of Commerce, Washington, DC, December 1982.
24. Medical Devices, Hearings Before the Subcommittee on Public Health and Environment, U.S. Congress House Interstate and Foreign Commerce, Serial No. 93–61, U.S. Government Printing Office, Washington, DC, 1973.
25. Murray, K., Canada's Medical Device Industry Faces Cost Pressures, Regulatory Reform, *Med. Dev. Diag. Ind. Mag.*, Vol. 19, No. 8, 1997, pp. 30–39.
26. Walter, C.W., Instrumentation Failure Fatalities, *Electronic News*, January 27, 1969.
27. Micco, L.A., Motivation for the Biomedical Instrument Manufacturer, *Proceedings of the Annual Reliability and Maintainability Symposium*, 1972, pp. 242–244.
28. Dhillon, B.S., Reliability Technology in Health Care Systems, *Proceedings of the IASTED International Symposium on Computers Advanced Technology in Medicine Health Care Bioengineering*, 1990, pp. 84–87.

29. Allen, R.C., FDA and the Cost of Health Care, *Med. Dev. Diag. Ind. Mag.*, Vol. 18, No. 7, 1996, pp. 28–35.
30. Schwartz, A.P., A Call for Real Added Value, *Medical Industry Executive*, February/March, 1994, pp. 5–9.
31. O'Leary, D.J., International Standards: Their New Role in Global Economy, *Proceedings of the Annual Reliability and Maintainability Symposium*, 1996, pp. 17–23.
32. Bethune, J., On Product Liability: Stupidity and Waste Abounding, *Med. Dev. Diag. Ind. Mag.*, Vol. 18, No. 8, 1996, pp. 8–11.
33. Federal Food, Drug, and Cosmetic Act, as Amended, Sec. 201(h), U.S. Government Printing Office, Washington, DC, 1993.
34. Onel, S., Draft Revision of FDA's Medical Device Software Policy Raises Warning Flags, *Med. Dev. Diag. Ind. Mag.*, Vol. 19, No. 10, 1997, pp. 82–91.
35. Naresky, J.J., Reliability Definitions, *IEEE Transactions on Reliability*, Vol. 19, 1970, pp. 198-200.
36. Omdahl, T.P., Ed., *Reliability, Availability and Maintainability (RAM) Dictionary*, ASQC Quality Press, Milwaukee, WI, 1988.
37. Von Alven, W.H., Ed., *Reliability Engineering*, Prentice-Hall, Inc., Englewood Cliffs, NJ, 1964.
38. Fries, R.C., *Reliable Design of Medical Devices,* Marcel Dekker, Inc., New York, 1997.
39. Mckenna, T., Oliverson, R., *Glossary of Reliability and Maintenance Terms*, Gulf Publishing Company, Houston, TX, 1997.
40. MIL-STD-721, Definitions of Effectiveness, Terms for Reliability, Maintainability, Human Factors, and Safety, Department of Defense, Washington, DC.

2 Basic Reliability Mathematics and Concepts for Medical Devices

CONTENTS

2.1 INTRODUCTION

Mathematical concepts play an important role in engineering, and the discipline of reliability relies heavily on probability concepts for its underlying support. Thus, for the real appreciation of reliability, a good understanding of probability theory is absolutely necessary. Originally, probability theory started with the study of games of chance. Two typical examples of these games are roulette and cards. With the passage of time, probability theory found applications in many other areas. A detailed history of probability is given in references 1 and 2.

The reliability discipline is based on many basic reliability concepts. Their understanding is essential to grasp advanced concepts of reliability. The starting point for the appreciation of these concepts is the clear understanding of definitions used in reliability work. The standard reliability networks such as series, parallel, k-out-of-n, and standby system are also important components of the basic reliability concepts. Needless to say, a good understanding of items such as these is vital for subsequent studies in reliability. Many of the basic reliability concepts are described in detail in reference 3.

By keeping the above factors in mind, this chapter presents useful basic reliability mathematics and concepts to understand subsequent chapters of the book.

The chapter also presents some fundamental nonmathematical aspects related to medical devices.

2.2 BASIC LAWS OF BOOLEAN ALGEBRA AND PROBABILITY

Boolean algebra is named after George Boole (1815–1864),[4] and it plays an important role in probability and reliability studies. There are many laws of Boolean algebra. Some of them are as follows:[4,5]

- Idempotent laws

$$X + X = X \tag{2.1}$$

$$XX = X \tag{2.2}$$

where

+ denotes the union of sets or events.
X is an arbitrary set or event.

The intersection of sets as in Equation (2.2) is written without the dot. However, in some books or other documents, it is written with a dot or with a symbol $\cap$.

- Absorption laws

$$X + XA = X \tag{2.3}$$

$$X(X + A) = X \tag{2.4}$$

where

A is an arbitrary set or event.

- Distributive laws

$$A + (XB) = (A + X)(A + B) \tag{2.5}$$

$$A(X + B) = AX + AB \tag{2.6}$$

where

B is an arbitrary set or event.

- Commutative laws

$$A + X = X + A \tag{2.7}$$

$$AX = XA \tag{2.8}$$

Reliability theory is based on the fundamental concepts of probability. There are many basic probability properties and some of them are as follows:[5,6]

- The probability of occurrence of event X is always

$$0 \leq P(X) \leq 1 \tag{2.9}$$

where

P(X) is the probability of occurrence of event X.

- The probability of occurrence and nonoccurrence of event X is always

$$P(X) + P(\overline{X}) = 1 \tag{2.10}$$

where

$P(\overline{X})$ is the probability of the nonoccurrence of event X.

Thus, from Equation (2.10) we get

$$P(X) = 1 - P(\overline{X}) \tag{2.11}$$

- The probability of an intersection of independent events $X_1, X_2, X_3, \ldots, X_m$ is given by

$$P(X_1 X_2 X_3 \ldots X_m) = \prod_{i=1}^{m} P(X_i) \tag{2.12}$$

where

m is the number of events.
$P(X_i)$ is the probability of occurrence of event X_i; for i = 1, 2, 3, …, m.

- The probability of the union of m mutually exclusive events is

$$P(X_1 + X_2 + X_3 + \ldots + X_m) = \sum_{i=1}^{m} P(X_i) \tag{2.13}$$

- The probability of the union of m independent events is expressed by

$$P(X_1 + X_2 + X_3 + \ldots + X_m) = 1 - \prod_{i=1}^{m} \left[1 - P(X_i)\right] \tag{2.14}$$

2.3 PROBABILITY DISTRIBUTIONS FOR TIME TO FAILURE

In reliability work, various types of statistical distributions are used to represent item times to failure.[7] This section presents four such distributions considered useful to perform reliability analysis of medical devices.

2.3.1 Exponential Distribution

This is probably the most widely used distribution to represent times to failure of engineering parts/systems, in particular the electronics.[8] More specifically, when the failure rate of engineering items is constant, their times to failure follow the exponential distribution. Also, this distribution is relatively easy to handle in conducting reliability analysis. Its probability density function is defined by

$$f(t) = \lambda e^{-\lambda t}, \; t \geq 0, \; \lambda > 0 \tag{2.15}$$

where

f(t) is the probability or failure density function.
t is time.
λ is the distribution parameter.

The cumulative distribution function, F(t), of a distribution is defined by

$$F(t) = \int_{-\infty}^{t} f(y)\, dy \tag{2.16}$$

Thus, using Equation (2.15) in Equation (2.16) yields

$$F(t) = 1 - e^{-\lambda t} \tag{2.17}$$

The above equation is the cumulative distribution for the exponential distribution.

2.3.2 Rayleigh Distribution

This distribution was developed by John Rayleigh (1842–1919)[2] and is often used in the theory of sound and in reliability studies, particularly in modeling a wear out characteristic.

When times to failure follow Rayleigh distribution, the failure or hazard rate increases linearly with time. Rayleigh probability density function is defined by

$$f(t) = \frac{2t}{\alpha^2} e^{-(t/\alpha)^2}, \quad t \geq 0, \alpha > 0 \tag{2.18}$$

where

α is the distribution parameter.

By inserting Equation (2.18) into Equation (2.16), we get the following expression for the Rayleigh cumulative distribution function:

$$F(t) = 1 - e^{-(t/\alpha)^2} \tag{2.19}$$

2.3.3 Weibull Distribution

This distribution was developed by a Swedish mechanical engineering professor, W. Weibull, working at the Royal Institute of Technology in Stockholm in the early 1950s.[9] Weibull distribution is useful to represent many different physical phenomena because it can model lifetimes having constant, increasing, and decreasing failure or hazard rate function.[10] Weibull probability or failure density function is defined by

$$f(t) = \frac{\beta t^{\beta-1} e^{-(t/\alpha)^\beta}}{\alpha^\beta}, \quad t \geq 0, \alpha > 0, \beta > 0 \tag{2.20}$$

where

α is the scale parameter.
β is the shape parameter.

Using Equation (2.20) in Equation (2.16) yields the following equation for the Weibull cumulative distribution function:

$$F(t) = 1 - e^{-(t/\alpha)^\beta} \tag{2.20a}$$

Noted that the exponential and Rayleigh distributions are the special cases of the Weibull distribution, i.e, for $\beta = 1$ and 2, respectively. For example, for $\beta = 1$ and $(1/\alpha) = \lambda$, Equation (2.20) is exactly the same as Equation (2.17). And for $\beta = 2$, it is identical to Equation (2.19).

2.3.4 General Distribution

This is a four parameter distribution, i.e., two scale and two shape parameters, and it is useful to model failure behavior of a wide range of items. The probability density of the distribution is defined by[11,12]

$$f(t) = \left[c\lambda u t^{u-1} + (1-c)\beta t^{\beta-1}\theta e^{\theta t^{\beta}}\right] \times \left[\exp\left[-c\lambda t^{u} - (1-c)\left(e^{\theta t^{\beta}} - 1\right)\right]\right] \quad \text{for} \quad \beta, \theta, u, \lambda > 0; 0 \le c \le 1; \quad \text{and} \quad t \ge 0 \tag{2.21}$$

where

u and β are the distribution shape parameters.
t is time.
λ and θ are the distribution scale parameters.

By substituting Equation (2.21) into Equation (2.16), we get the following equation for the cumulative distribution function of the general distribution:

$$F(t) = 1 - \exp\left[-c\lambda t^{u} - (1-c)\left(e^{\theta t^{\beta}} - 1\right)\right] \tag{2.22}$$

In special cases, the general distribution becomes

- Bathtub hazard rate curve (i.e., for $u = 0.5$, $\beta = 1$)
- Extreme value (i.e., for $c = 0$, $\beta = 1$)
- Makeham (i.e., for $u = 1$, $\beta = 1$)
- Weibull (i.e., for $c = 1$)
- Rayleigh (i.e., for $u = 2$, $c = 1$)
- Exponential (i.e., for $c = 1$, $u = 1$)

All in all, the general distribution is quite useful to represent the bathtub hazard or failure rate curve associated with many engineering items.

2.4 BASIC RELIABILITY-RELATED DEFINITIONS

There are many definitions used in performing reliability analysis of engineering items, and their clear understanding is important to effectively conduct such analysis. This section presents basic definitions of reliability, mean time to failure, and hazard rate.

2.4.1 Reliability

This is defined by

$$R(t) = 1 - F(t)$$
$$= 1 - \int_{-\infty}^{t} f(x)dx \tag{2.23}$$

or

$$R(t) = \int_{t}^{\infty} f(x)dx \tag{2.24}$$

or

$$R(t) = e^{-\int_0^t \lambda(t)dt} \tag{2.25}$$

where

R(t) is the reliability at time *t*.
λ(t) is hazard rate or time dependent failure rate.

It means Equation (2.23), (2.24), or (2.25) can be used to obtain an expression for reliability of an item.

Example 2.1

Assume that the times to failure of a respirator are described by the following probability density function.

$$f(t) = \lambda e^{-\lambda t},\ \lambda > 0,\ t \geq 0 \tag{2.26}$$

where

t is time.
λ is the respirator failure rate.

Obtain an expression for the respirator reliability by using Equations (2.23) and (2.24). By substituting Equation (2.26) into Equation (2.23), we get

$$R(t) = 1 - \int_0^t \lambda e^{-\lambda x} dx$$
$$R(t) = e^{-\lambda t} \tag{2.27}$$

Similarly, inserting Equation (2.26) into Equation (2.24) yields

$$R(t) = \int_t^{\infty} \lambda e^{-\lambda x} dx = \left[\frac{\lambda e^{-\lambda x}}{-\lambda}\right]_t^{\infty} = e^{-\lambda t} \quad (2.28)$$

As Equations (2.27) and (2.28) are identical, it proves that the application of Equations (2.23) and (2.24) gives exactly the same result.

2.4.2 Mean Time to Failure

This is defined by

$$MTTF = \int_0^{\infty} R(t) dt \quad (2.29)$$

or

$$MTTF = \int_0^{\infty} t f(t) dt \quad (2.30)$$

or

$$MTTF = \lim_{s \to 0} R(s) \quad (2.31)$$

where

MTTF is the mean time to failure.
s is the Laplace transform variable.
R(s) is the Laplace transform of the reliability function, R(t). The Laplace transform of a function f(t) is defined by

$$f(s) = \int_0^{\infty} f(t) e^{-st} dt, \quad (2.32)$$

where

f(s) is the Laplace transform of f(t).

Example 2.2

Assume that the reliability function of a cardiac defibrillator is

$$R(t) = e^{-\lambda t} \tag{2.33}$$

where

λ = 0.002 failures/hour.
t is time.

Find the defibrillator mean time to failure.

Using Equation (2.33) in Equation (2.29) yields

$$\begin{aligned} MTTF &= \int_0^\infty e^{-(0.002)t} dt \\ &= 1/0.002 \\ &= 500 \text{ hours} \end{aligned}$$

The cardiac defibrillator mean time to failure is 500 hours.

2.4.3 Hazard Rate

This is defined by

$$\lambda(t) = \frac{f(t)}{R(t)} \tag{2.34}$$

or

$$\lambda(t) = \frac{f(t)}{1 - F(t)} \tag{2.35}$$

or

$$\lambda(t) = -\frac{1}{R(t)} \frac{dR(t)}{dt} \tag{2.36}$$

where

$\lambda(t)$ is hazard rate or time dependent failure rate.

It means any of the above three equations can be used to obtain the hazard rate of an item.

Example 2.3

In example 2.1, obtain an expression for the respirator hazard rate. Comment on the result.

By inserting Equation (2.27) into Equation (2.36), we get

$$\lambda(t) = -\frac{1}{e^{-\lambda t}} \frac{de^{-\lambda t}}{dt} \quad (2.37)$$

$$= \lambda$$

It means the hazard rate of the respirator is constant. More specifically, it does not depend on time.

2.5 RELIABILITY CONFIGURATIONS

In reliability evaluation studies, the engineering systems may form various types of configurations. This section presents reliability evaluation of some commonly occurring configurations: series, parallel, k-out-of-n, and standby system.

2.5.1 Series Configuration

This is the simplest and probably the most commonly occurring or assumed configuration in reliability evaluation of engineering systems. The success of the series system depends on the success of all its elements. More specifically, if any of the elements fail, the system fails. Two typical examples of this type of systems are a series string of Christmas tree lights and four car tires. Figure 2.1 shows a block diagram of a series system. Each block in the diagram represents a system element or unit.

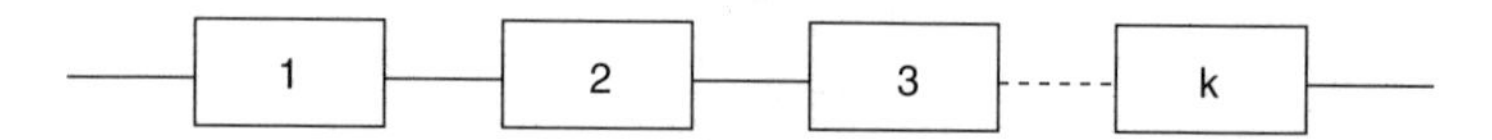

FIGURE 2.1 Block diagram of k unit series system.

For independent units, the reliability of the series system shown in Figure 2.1 is given by

$$R_s = R_1 R_2 R_3 \ldots R_k \quad (2.38)$$

where

R_s is the series system reliability.
k is the number of independent units in series.
R_i is the unit i reliability; for i = 1,2,3,…k.

For constant failure rate of unit i, i.e., $\lambda_i(t) = \lambda_i$, from Equation (2.25) we get

$$R_i(t) = e^{-\lambda_i t} \tag{2.39}$$

where

$R_i(t)$ is the reliability of unit i at time t.

Substituting Equation (2.39) into Equation (2.38) yields the following expression for the series system time dependent reliability:

$$R_s(t) = e^{-\sum_{i=1}^{k} \lambda_i t}, \tag{2.40}$$

where

$R_s(t)$ is the series system reliability at time t.

Using Equation (2.40) in Equation (2.29) yields

$$MTTF_s = \frac{1}{\sum_{i=1}^{k} \lambda_i}, \tag{2.41}$$

where

$MTTF_s$ is the series system mean time to failure.

By substituting Equation (2.40) into Equation(2.36), we obtain

$$\lambda_s(t) = \sum_{i=1}^{k} \lambda_i, \tag{2.42}$$

where

$\lambda_s(t)$ is the series system hazard or failure rate.

Note that the right side is independent of time *t*. It means the series system failure rate is simply the sum of the unit failure rates.

Example 2.4

Assume that a magnetic resonance imaging (MRI) machine is composed of five identical subsystems in series. Subsystems 1, 2, 3, 4, and 5 constant failure rates are $\lambda_1 = 0.0004$ failures/hour, $\lambda_2 = 0.0005$ failures/hour, $\lambda_3 = 0.0006$ failures/hour, $\lambda_4 = 0.0007$ failures/hour, and $\lambda_5 = 0.0001$ failures/hour, respectively. Calculate the MRI machine mean time to failure.

Substituting the given data into Equation (2.41) yields

$$MTTF_s = \frac{1}{0.0004 + 0.0005 + 0.0006 + 0.0007 + 0.0001}$$
$$= 434.78 \text{ hours}$$

The mean time to failure of the MRI machine is 434.78 hours.

2.5.2 Parallel Configuration

In this case, all units are active and at least one unit must perform successfully for the system success. Figure 2.2 shows the block diagram of a parallel system, and each block in the diagram represents a unit. This type of arrangement is used to improve reliability of a system. In the parallel system shown in Figure 2.2, reliability is expressed by

$$R_p = 1 - (1 - R_1)(1 - R_2)\ldots(1 - R_n), \tag{2.43}$$

where

R_p is the parallel system reliability.
n is the number of independent units.
R_i is the unit i reliability; for i = 1, 2, 3, …, n.

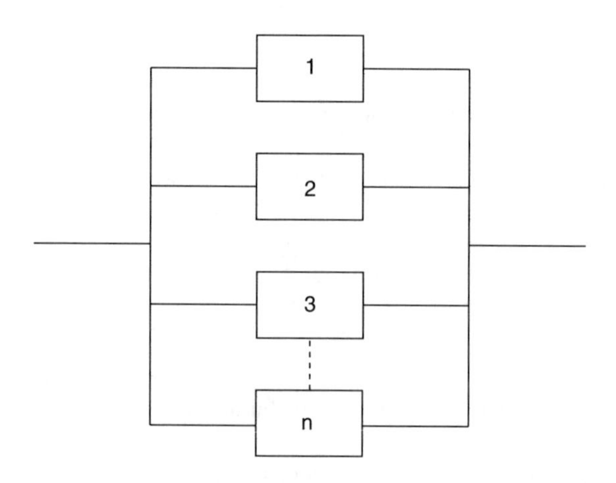

FIGURE 2.2 An n unit parallel system.

For constant failure rates of parallel units, substituting Equation (2.39) into Equation (2.43) yields

$$R_p(t) = 1 - \left(1 - e^{-\lambda_1 t}\right)\left(1 - e^{-\lambda_2 t}\right)\ldots\left(1 - e^{-\lambda_n t}\right), \tag{2.44}$$

where

$R_p(t)$ is the parallel system reliability at time t.
λ_i is the constant failure rate of unit i; for i = 1, 2, …, n.

For identical units, $\lambda_i = \lambda$, Equation (2.44) becomes

$$R_p(t) = 1 - (1 - e^{-\lambda t})^n \quad (2.45)$$

By substituting Equation (2.45) into Equation(2.29), we get

$$MTTF_p = \frac{1}{\lambda} \sum_{i=1}^{n} 1/i, \quad (2.46)$$

where

$MTTF_p$ is the parallel system mean time to failure.

Example 2.5

A medical system is composed of two independent and identical subsystems in parallel, and at least one subsystem must operate normally for the system success. The failure rate of a subsystem is 0.0001 failures per hour. Calculate the medical system reliability for a 100-hour mission and its mean time to failure. Compare the system mean time to failure with that of a subsystem.

Inserting the given data into Equation (2.45) yields

$$R_p(10) = 1 - \left(1 - e^{-(0.0001)(100)}\right)^2$$
$$= 0.9999$$

Using the specified data in Equation (2.46), we obtain

$$MTTF_p = \frac{1}{(0.0001)} \frac{3}{2}$$
$$= 15{,}000 \text{ hours}$$

The mean time to failure of a single subsystem is

$$MTTF = \frac{1}{\lambda}$$
$$= 10{,}000 \text{ hours}$$

It means that by putting two subsystems in parallel, we were able to increase mean time to failure from 10,000 hours to 15,000 hours.

2.5.3 K-OUT-OF-N CONFIGURATION

This is basically the same as the parallel configuration but instead of at least one unit, k units out of a total of n units must work normally for the system success. For independent and identical units, the reliability of the k-out-of-n configuration is

$$R_{k/n} = \sum_{i=1}^{n} \binom{n}{i} R^{i} (1-R)^{n-i}, \tag{2.47}$$

where

$$\binom{n}{i} = \frac{n!}{i!(n-i)!}$$

$R_{k/n}$ is the reliability of the k-out-of-n system.
R is the unit reliability.

For constant unit failure rate, Equation (2.47) becomes

$$R_{k/n}(t) = \sum_{i=k}^{n} \binom{n}{i} e^{-i\lambda t} \left(1 - e^{-\lambda t}\right)^{n-i} \tag{2.48}$$

Using Equation (2.48) in Equation (2.29) yields

$$MTTF_{k/n} = \frac{1}{\lambda} \sum_{i=k}^{n} 1/i \tag{2.49}$$

where

$MTTF_{k/n}$ is the mean time to failure of the k-out-of-n system.

For k = 1 and k = n, the k-out-of-n system becomes parallel and series systems, respectively.

Example 2.6

Assume that a subsystem of an X-ray machine is made up of three independent and identical units in parallel and at least two units must operate normally for the subsystem success. Each unit's probability of success is 0.8. Calculate the subsystem reliability.

By substituting the given values into Equation (2.47), we get

$$R_{2/3} = \sum_{i=2}^{3} \binom{3}{i}(0.8)^i (1-0.8)^{3-i}$$

$$= \binom{3}{2}(0.8)^2(1-0.8) + \binom{3}{3}(0.8)^3(1-0.8)^0$$

$$= 0.8960$$

It means, there is approximately a 90% chance that the X-ray machine's subsystem will not fail.

2.5.4 Standby System

This is another form of redundancy used to improve system reliability. In this case, only one unit operates and the remaining units are kept as standbys. As soon as the operating unit fails, it is immediately replaced with one of the standbys. Figure 2.3 shows a block diagram of a standby system. Each block in the diagram denotes a unit. In this diagram, one unit is operating and n units are on standby. It means the system has a total of (n + 1) units.

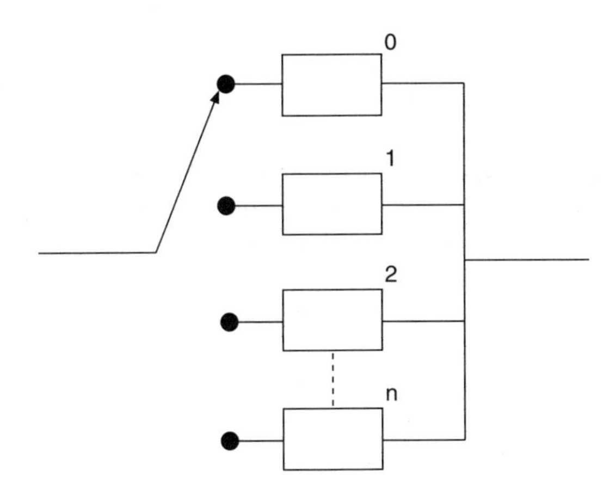

FIGURE 2.3 Block diagram of a (n + 1) unit standby system.

The reliability of the standby system is given by

$$R_{ss}(t) = \left\{ \frac{\sum_{i=0}^{n} \left[\int_0^t \lambda(t)dt \right]^i e^{-\int_0^t \lambda(t)dt}}{i!} \right\}, \tag{2.50}$$

where

$R_{ss}(t)$ is the standby system reliability.

Equation (2.50) is subject to the following assumptions:

- Independent and identical units.
- Perfect switching mechanism.
- Standby units remain as good as new in their standby mode.
- Time dependent unit failure rate.

For constant unit failure rate, λ, Equation (2.50) becomes

$$Rss(t) = \sum_{i=0}^{n} (\lambda t)^i e^{-\lambda t}/i! \tag{2.51}$$

By substituting Equation (2.51) into Equation (2.29), we obtain

$$MTTF_{ss} = \frac{n+1}{\lambda} \tag{2.52}$$

Example 2.7

Assume that a medical system is composed of two independent and identical units in a standby arrangement, i.e., one unit operating and the other on standby. Also, the switching mechanism is perfect and the unit failure rate is 0.005 failures/hour. Calculate the medical system reliability for a 50-hour mission.

By substituting the given data into Equation (2.51), we get

$$R_{ss}(50) = e^{-(0.005)(50)}\left[1 + (0.005)(50)\right]$$
$$= 0.9735$$

The medical system reliability is 0.9735.

2.6 AEROSPACE AND MEDICAL EQUIPMENT RELIABILITY COMPARISONS

The development of the reliability field is basically associated with the development of aerospace equipment, and its history may be traced back to World War II, with the design and development of V1 and V2 rockets by the Germans. In comparison, the application of reliability concepts to medical equipment may be traced back to the latter part of the 1960s.[13-17] Nonetheless, some comparisons between aerospace equipment reliability and medical device reliability are presented in Table 2.1.[18-19]

TABLE 2.1
Comparisons Between Aerospace Equipment Reliability and Medical Reliability

No.	Aerospace Equipment Reliability	Medical Device Reliability
1	Costly equipment.	Relatively less costly equipment.
2	A well-established reliability field.	A relatively new area for the application of reliability principles.
3	Large manufacturing organizations with an extensive reliability-related experience and well established reliability engineering departments.	Relatively small manufacturing organizations with much less reliability-related experience and less established reliability engineering departments.
4	Reliability professionals with extensive related past experience who use sophisticated reliability approaches.	Reliability professionals with relatively less related past experience who employ simple reliability methods.
5	Well-being of humans is a factor directly or indirectly.	Lives or well-being of patients are involved.

2.7 GUIDELINES FOR RELIABILITY PROFESSIONALS WORKING IN HEALTH CARE

Today many reliability professionals work in the medical device industry. Over the years, many guidelines for such professionals have been developed to improve device reliability. Some of those guidelines are as follows:[18-19]

- Direct your attention to device critical failures, as not all device failures are equal in importance.
- Remember that approaches such as failure modes and effect analysis (FMEA), fault tree analysis (FTA), design review and parts review are useful to obtain immediate results.
- Train your mind to be a cost-conscious reliability professional because management usually approves only those reliability-related decisions that are considered cost-effective.
- Use simple and straightforward reliability methods and techniques. Seriously consider using methods such as FMEA and qualitative FTA. Remember that the responsibility of device reliability rests with the manufacturer during its design and maufacture, and in the field use, it is generally the responsibility of the user.
- Past experience has shown that a simple failure reporting system is generally more beneficial in terms of improving medical device reliability rather than having a large number of spares or standbys.

2.8 PROBLEMS

1. Describe the following laws associated with the Boolean algebra:
   - Absorption laws
   - Idempotent laws
2. Prove the following

$$A + (XB) = (A + X)(A + B), \tag{2.53}$$

   where
   A is an arbitrary set or event.
   B is an arbitrary set or event.
   X is an arbitrary set or event.
3. Write down three approaches for obtaining mean time to failure of an item.
4. Prove in Question 3 that all three mean time to failure methods give the same result.
5. Obtain a mean time to failure expression for the Rayleigh distribution.
6. Prove that

$$R(t) + F(t) = 1, \tag{2.54}$$

   where
   R(t) is the reliability at time *t*.
   F(t) is the failure probability at time *t*.
7. Assume that the times to failure of a spectrophotometer are Weibull distributed. Obtain an expression for the spectrophotometer hazard rate.
8. Make a comparison of aerospace and medical equipment reliability.
9. Assume that a medical system is composed of three independent and identical units in a standby arrangement, i.e., one unit operating, the other two on standby. The switching mechanism is perfect, and the unit failure rate is 0.0001 failures/hour. Calculate the medical system reliability for a 100-hour mission.
10. A medical system is composed of two independent units in parallel. The failure rate of unit 1 is 0.004 failures/hour and that of unit 2 is 0.007 failures/hour. Calculate the system mean time to failure.

REFERENCES

1. Owen, D.B., Ed., *On the History of Statistics and Probability*, Marcel Dekker Inc., New York, 1976.
2. Eves, H., *An Introduction to the History of Mathematics*, Holt, Rinehart, and Winston, New York , 1976.
3. Dhillon, B.S., *Systems Reliability, Maintainability, and Management*, Petrocelli Books Inc., New York, 1983.

4. Lipschutz, S., *Set Theory and Related Topics*, McGraw-Hill Book Company, New York, 1964.
5. Lipschutz, S., *Probability*, McGraw-Hill Book Company, New York, 1965.
6. Shooman, M.L., *Probabilistic Reliability: An Engineering Approach*, McGraw-Hill Book Company, New York, 1968.
7. Dhillon, B.S., *Reliability Engineering in Systems Design and Operation*, Van Nostrand Reinhold Company, New York, 1983.
8. Davis, D.J., An Analysis of Some Failure Data, *J. Am. Stat. Assoc.*, June 1952, pp. 113–150.
9. Weibull, W., A Statistical Distribution Function of Wide Applicability, *J. Appl. Mech.*, Vol. 18, 1951, pp. 293–297.
10. Leeemis, L.M., *Reliability: Probabilistic Models and Statistical Methods*, Prentice-Hall Inc., Englewood Cliffs, NJ, 1995.
11. Dhillon, B.S., A Hazard Rate Model, *IEEE Transaction on Reliability*, Vol. 29, 1979, p. 150.
12. Dhillon, B.S., Life Distributions, *IEEE Transaction on Reliability*, Vol. 30, 1981, pp. 457–460.
13. Johnson, J.P., Reliability of ECG Instrumentation in a Hospital, *Proceedings of the Annual Symposium on Reliability*, 1967, pp. 314–318.
14. Meyer, J.L., Some Instrument Induced Errors in the Electrocardiogram, *J. Am. Med. Assoc.*, Vol. 201, 1967, pp. 351–358.
15. Gechman, R., Tiny Flaws in Medical Design Can Kill, *Hop. Top.*, 1968, pp. 23–24.
16. Crump, J.F., Safety and Reliability in Medical Electronics, *Proceedings of the Annual Symposium on Reliability*, 1969, pp. 320–322.
17. Taylor, E.F., The effect of Medical Test Instrument Reliability on Patient Risks, *Proceedings of the Annual Symposium on Reliability*, 1969, pp. 328–330.
18. Dhillon, B.S., *Reliability Engineering in Systems Design and Operation*, Van Nostrand Reinhold Company, New York, 1983.
19. Taylor, E.F., The Reliability Engineer in the Health Care System, *Proceedings of the Annual Reliability and Maintainability Symposium*, 1972, pp. 245–248.

3 Tools for Medical Device Reliability Assurance

CONTENTS

3.1 INTRODUCTION

There are many methods and techniques used to assure reliability of engineering systems or items. Many of these methods and techniques can also be used to assure reliability of medical devices. Therefore, this chapter presents such methods as: failure mode and effects analysis (FMEA), fault tree analysis (FTA), failure rate evaluation and parts count methods, Markov method, and the common-cause failure analysis method.

FMEA is a powerful design tool to analyze engineering systems. It was developed in the early 1950s to analyze flight control systems from the reliability aspects.[1,2] FTA is one of the most widely used methods in the industrial sector to perform failure analysis of engineering systems, especially in the nuclear power generation. Originally, the fault tree approach was developed to perform analysis of the Minuteman Launch Control System in the early 1960s.[3]

The failure rate evaluation and parts count methods are practically inclined. In particular, the parts count approach is quite useful during bid proposal in estimating equipment/item failure rate during early design phases. The method first appeared in a document entitled "Reliability Prediction of Electronic Equipment."[4]

The Markov method is a basic approach and can generally handle more cases than any other method, in particular, the repairable systems. The method is named after a Russian mathematician. The common-cause failure analysis method is basically a block diagram approach used to analyze systems with common-cause failures. It was developed during the middle of the 1970s.[5,6] A common-cause failure is any instance where multiple units or components fail due to a single cause.

This chapter describes in detail all of the above methods.

3.2 FAILURE MODE AND EFFECT ANALYSIS (FMEA)

This is a bottom-up approach, and its objective is to improve items/system reliability during the design phase by identifying all conceivable and potential failure modes and determining the effect of each on system/item performance. FMEA is known as failure mode effects and criticality analysis (FMECA) when critical analysis (CA) is added to it.[7-9] CA may simply be described as a quantitative approach used to rank critical failure mode effects by considering their occurrence probability.

A comprehensive list of publications on FMEA is given in reference 10.

3.2.1 Important Terms and Definitions

As there are many terms and definitions involved to perform FMEA/FMECA, this section presents some of the important ones:[7]

- **Failure mode**. This is the manner through which a failure is observed. More specifically, it usually describes the way the failure occurs and its effect on item/system operation.
- **Criticality**. This is a relative measure of a failure mode consequences and its occurrence frequency.
- **Failure cause**. This is design defects, physical or chemical processes, part misapplications, quality defects, or other processes that are the primary reason for failure or which start the physical process through which deterioration leads to malfunction or failure.
- **Failure effect**. This is the consequence or consequences of a failure mode on the operation, function, or status of an item/system.
- **Single failure point**. This is the item failure that will lead to system failure and is not compensated for by redundancy or alternative operational approach or procedure.

3.2.2 FMEA Procedure

This procedure is accomplished through formal documentation because of its benefits such as to standardize the procedure, a means of historical documentation, and a basis for future improvements. Nonetheless, the FMEA procedure is composed of a sequence of eight logical steps as shown in Figure 3.1.[7]

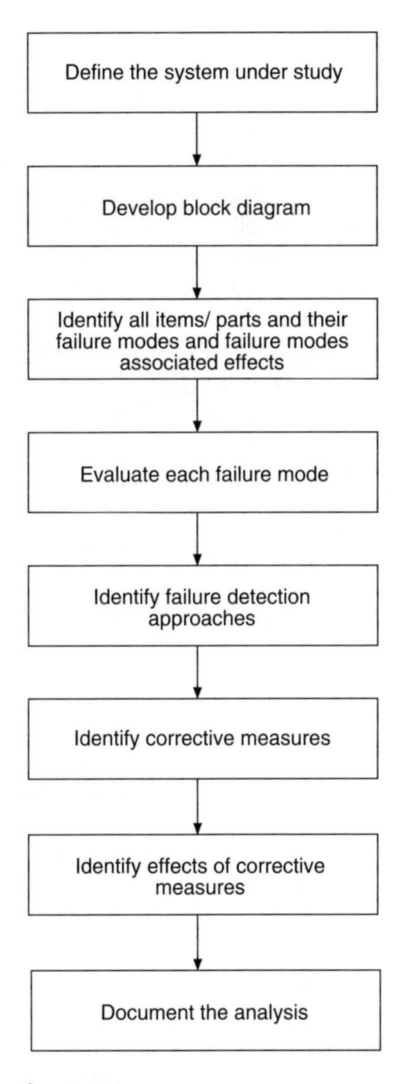

FIGURE 3.1 FMEA performance steps.

The first step is concerned with defining the system under consideration. This includes identifying internal and interface functions, failure definitions, system restraints, and expected performance at all indenture levels. Furthermore, the system functional narratives should include each mission's description in terms of functions, which highlights tasks to be conducted for each mission, operational mode, and mission phase. Also, the narratives should describe items such as environment profiles, each part's functions and outputs, and expected mission times and equipment utilization.

Step 2 is concerned with developing block diagrams. This requires for each item configuration involved in the system's use developing functional and reliability block diagrams illustrating the operation, interrelationships, and interdependencies of functional entities.

The third step is concerned with the identification of all items/parts and their failure modes and effects associated with the failure mode. More specifically, in this step all potential item and interface failure modes are identified and their effect(s) on the immediate function/item/system/mission defined.

Step 4 concerns evaluating each failure mode in terms of the worst case scenario potential consequences as well as assigning a severity classification category.

In step 5, failure detection methods are identified along with a compensating provisions for each failure mode.

Step 6 involves the identification of corrective design or other measures necessary to eradicate the failure or control the risk.

In step 7, the effects of corrective measures or other system attributes such as requirement for logistics support are identified.

Step 8, the final step of the FMEA procedure, concerns the documentation of the analysis. In addition, the problems that cannot be rectified through design are summarized, and the special controls appropriate to reduce failure risk are identified.

3.2.3 FMEA Worksheet and Critical Item List

Normally a worksheet is used to perform FMEA. Even though its design may vary from one organization to another, Figure 3.2 presents a typical FMEA worksheet.[7,9] A detailed description of the FMEA worksheet is given in reference 7.

The critical item list is extracted from the FMEA and is compiled to provide input to sound management decisions. Usually, a critical item list contains information on areas such as the following:[7]

- Identification of critical item and FMEA cross reference
- Item failure mode
- Loss effect
- Classification of criticality
- Design features that help reduce failure occurrence probability for the item under consideration
- Tests performed to verify design features and tests planned at hardware acceptance/during field, which would help detect the occurrence of failure mode
- Planned inspections to ensure hardware is being built per the design requirements, and inspections planned during down time or maintenance that could detect the occurrence of failure mode or the conditions that could lead to the failure mode
- Statement that relates to the history of the design under consideration or the similar design
- Method through which the failure mode occurrence is detected
- Rationale for not attempting to eradicate failure mode(s)

Item identification numbers	Item / functional identification (nomenclature)	Item function	Operational mode or mission phase	Item failure modes and causes	Failure effects			Method for failure detection	Compensating provisions	Severity class or category	Analyst's remarks
					Local effects	Next higher level	End effects				
System: ____ Subsystem: ____ Part: ____ Drawing: ____					Failure Mode and Effects Analysis			By: ____ Page: ____ of ____ Date: ____			

FIGURE 3.2 A typical FMEA worksheet.

3.2.4 FMEA Benefits

There are many advantages associated with performing FMEA; some of those are

- Easy to comprehend
- A systematic approach to classify failures
- A useful tool to reduce engineering changes, development time, and cost
- Useful to analyze small, large, and complex systems
- Useful to compare designs
- A visibility tool for managers
- Useful to improve communication among design interface personnel
- A method that begins from the detailed level and works upward

3.3 FAULT TREE ANALYSIS (FTA)

FTA is a top-down approach of system analysis that is used to determine the possible occurrence of undesirable events or failures. Over the years, the method has gained favor over other reliability analysis approaches because of its versatility in degree of detail of complex systems. The technique was developed in 1961 by H.A. Watson of Bell Telephone Laboratories to analyze the Minuteman Missile Launch Control System.[11]

FTA is a detailed deductive analysis that generally needs a considerable amount of information about the system under study and ensures that all critical aspects of the system are identified and subsequently controlled. Furthermore, a fault tree may be described as a graphical representation of Boolean logic concerning the development of a particular undesirable system event or failure called the top event to basic failures known as primary events. Some examples of the top event could be "X-ray machine failure," "release of radioactivity from a nuclear power generating station," and "magnetic resonance imaging (MRI) machine failure."

Nonetheless, FTA begins by identifying a top event, such as listed above, associated with a system. Fault events, which can cause the top event to occur, are generated and connected by logic operators or gates such as OR and AND. The OR gate provides a true (failure) output when one or more of its inputs are true (failures), and AND gate provides a time (failure) output if all of its inputs are true (failures). Nonetheless, the fault tree construction proceeds by generation of fault events in a successive manner until the time when the fault events do not need to be developed further.

3.3.1 FAULT TREE SYMBOLS AND STEPS FOR FAULT TREE ANALYSIS

There are many symbols used to construct fault trees.[12] The basic four symbols are shown in Figure 3.3. The definitions of each of these symbols are

- **OR gate**. It denotes that an output fault event occurs if one or more of input fault event occur.
- **AND gate**. It denotes that an output fault tree event occurs if all the input fault events occur.
- **Rectangle**. It denotes a fault event that occurs from the logical combination of fault events through the input of logic gates such as OR and AND.
- **Circle**. It denotes a basic fault event or the failure of an elementary system component. The event's occurrence probability, failure, and repair rates are usually obtained from empirical data or other sources.

FTA can be conducted in a number of steps. In fact, in the published literature these steps may vary quite considerably. However, we propose the following steps:[8]

- Define undesirable or top event to be investigated.
- Establish boundaries.
- Thoroughly understand the system under consideration.

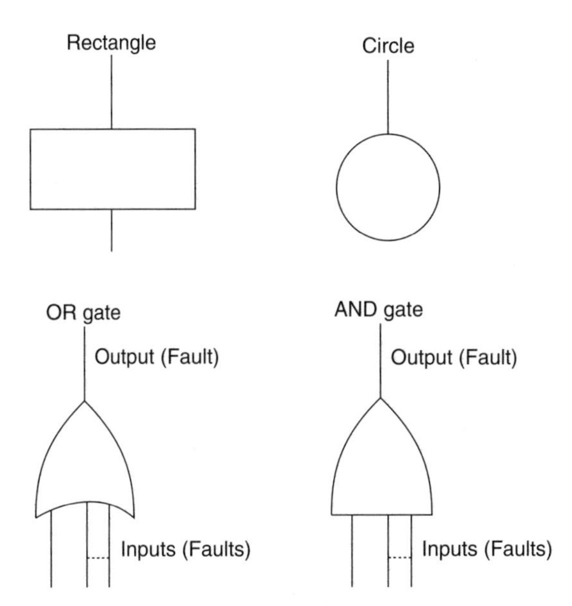

FIGURE 3.3 Basic fault tree symbols.

- Construct fault tree for the system under study.
- Perform analysis of the constructed fault tree.
- Take corrective measures as appropriate.

3.3.2 Probability Evaluation of OR and AND Gates

OR Gate

An n input fault events x_i, for i = 1, 2, ..., n, OR gate is shown in Figure 3.4. The probability of the occurrence of output fault X is expressed by

$$P(X) = 1 - \prod_{i=1}^{n} \{1 - P(x_i)\} \tag{3.1}$$

where

n is the total number of independent fault events.
x_i is the input fault event i; for i = 1, 2, 3, ..., n.
$P(x_i)$ is the probability of occurrence of event x_i.

AND Gate

Figure 3.5 shows an n input fault events x_i, for i = 1, 2, 3, ..., n AND gate. With the aid of Figure 3.5 diagram and for independent input fault events, the probability of occurrence of output fault event Y is given by

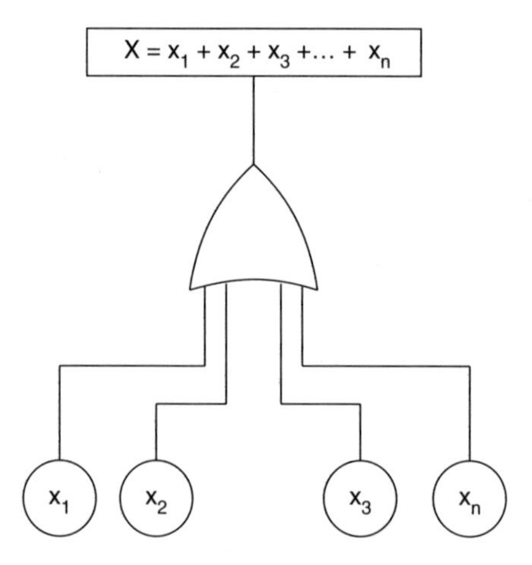

FIGURE 3.4 An OR gate with n input faults.

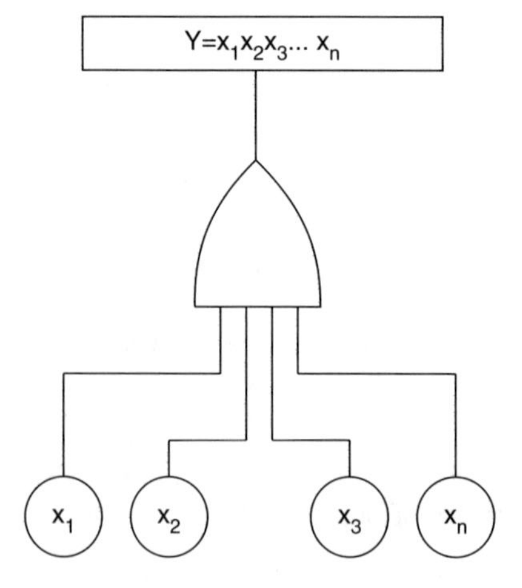

FIGURE 3.5 An AND gate with n input faults.

$$P(Y) = \prod_{i=1}^{n} P(x_i) \tag{3.2}$$

Example 3.1

Assume that a piece of medical equipment can be represented by a block diagram shown in Figure 3.6. Each block in the diagram represents the medical equipment part, and in turn, each part's failure event is denoted by c_i, for i = 1, 2, 3, 4, and 5. The failure probabilities of all the parts are also given in Figure 3.6. Develop a fault tree for the medical equipment failure by using Figure 3.6. Assume that the top event

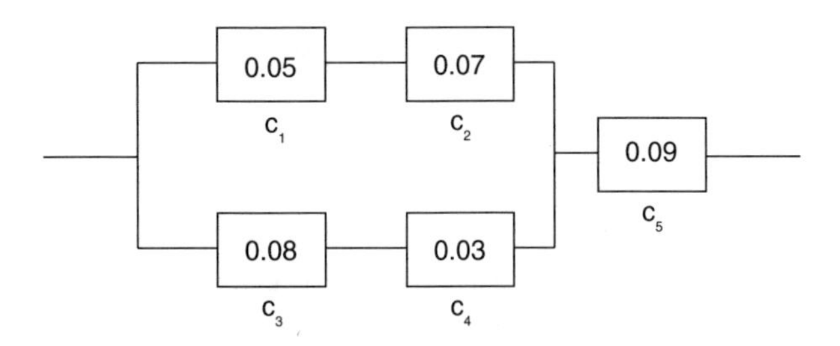

FIGURE 3.6 Block diagram of a piece of medical equipment.

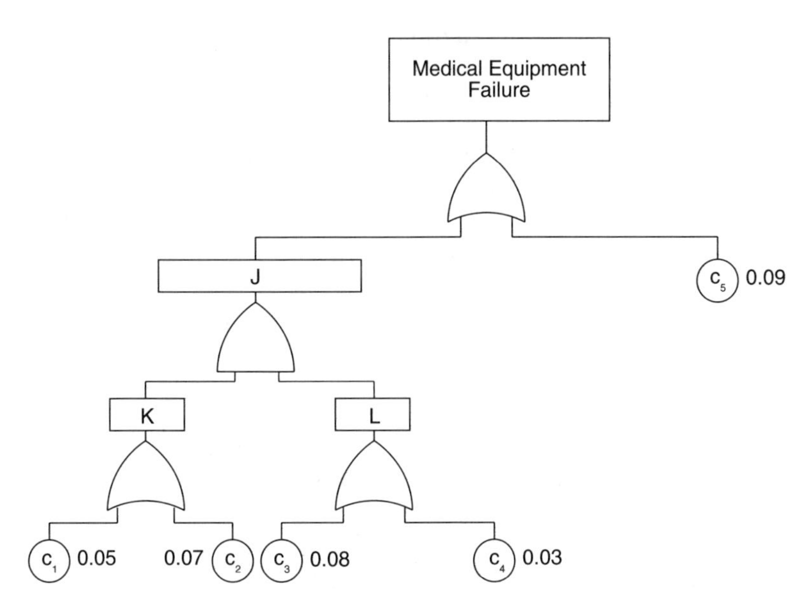

FIGURE 3.7 Fault tree for the top event "medical equipment failure."

is "medical equipment failure" and all the equipment parts fail independently. Calculate the probability of occurrence of the top event.

Fault tree for the block diagram of Figure 3.6 is shown in Figure 3.7. The figure shows the basic events and their occurrence probabilities, the intermediate events, (i.e., J, K, and L) and top event. Thus, using Figure 3.7 and Equation (3.1), the probability of occurrence of event K is

$$
\begin{aligned}
P(K) &= P(c_1) + P(c_2) - P(c_1)P(c_2) \\
&= 0.05 + 0.07 - (0.05)(0.07) \\
&= 0.1165.
\end{aligned}
$$

Similarly, the probability of occurrence of event L is

$$P(L) = P(c_3) + P(c_4) - P(c_3)P(c_4)$$
$$= 0.08 + 0.03 - (0.08)(0.03)$$
$$= 0.1076.$$

Using Figure 3.7, Equation (3.2), and the above calculations, the probability of occurrence of event J is given by

$$P(J) = P(K)P(L)$$
$$= (0.1165)(0.1076)$$
$$= 0.0125.$$

Using the above result and the given data for event c_5 in Equation (3.1), the probability of the medical equipment failure is given by

$$P_{te} = P(J) + P(c_5) - P(J)P(c_5)$$
$$= 0.0125 + 0.09 - (0.0125)(0.09)$$
$$= 0.1014,$$

where P_{te} is the probability of occurrence of the top event, i.e., medical equipment failure. It means, there is approximately a 10% chance that the medical equipment will fail.

3.3.3 Advantages and Disadvantages of the FTA

Just like any other reliability analysis technique, the FTA has its advantages and disadvantages. Some of the advantages include a visibility tool, deductive identification of failures, and relatively easy to handle complex systems. Analysts must understand the system thoroughly and deal specifically with one particular failure at a time. The reliability analysis could be either qualitative or quantitative.

Disadvantages of the FTA are that it is costly, time consuming, difficult to check results, and difficult to handle components with partial failure states.

3.4 FAILURE RATE EVALUATION AND PARTS COUNT METHODS

The failure rate evaluation method basically concerns evaluating the failure rates of electronic parts. It is often used during the design phase of an engineering item. Most of the failure rate evaluation models for electronic parts are given in MIL-HDBK-217.[4] Although the part failure rate models vary with different part types, their general form is as follows:[4,13]

$$\lambda_p = \lambda_b \, \alpha_e \, \alpha_a \, \alpha_q \ldots \alpha_n \tag{3.3}$$

where

- λ_p is the part failure rate.
- λ_b is the base failure rate. It is usually expressed by a model that relates the influence of electrical and temperature stresses on the part.
- α_e is the environmental adjustment factor that takes into consideration the influence of environments other than temperature. More specifically, it is related to the operating condition such as vibration and humidity under which the item must perform.
- α_a is the application adjustment factor whose value depends on the part application. More specifically, this factor takes into consideration "reliability significant" secondary stress and application factors.
- α_q is the quality adjustment factor, which takes into account the degree of manufacturing control under which the part was fabricated and tested.
- α_n is the symbol for other adjustment factors that take into account cyclic effects, construction class, etc.

The values of the above factors and λ_b, are given in references 4 and 13. The following equation is used to determine the base failure rate of many electronic parts:

$$\lambda_b = C \exp[-AE/kT] \tag{3.4}$$

where

- C is a constant.
- T is the absolute temperature.
- AE is the activation energy for the process.
- k is the Boltzmann's Constant.

Example 3.2

A failure rate equation for connections from reference 4 is

$$\lambda_c = \lambda_b \alpha_e \text{ failures}/10^6 \text{ hours}, \tag{3.5}$$

where

- λ_c is the connection failure rate.

As per reference 4, the values of the base failure rate and the environmental adjustment factor are as follows, respectively:

$$\lambda_b = 0.00026 \text{ failures}/10^6 \text{ (Connection type: crimp)}$$

$$\alpha_e = 2.0 \text{ (use environment: ground, fixed)}$$

Calculate crimp connection's failure rate.

Using the specified data in Equation (3.5), we get

$$\lambda_c = (0.00026)(2)$$

$$= 0.00052 \text{ failures}/10^6 \text{ hours}$$

Thus, the failure rate of the crimp connection is 0.00052 failures/10^6 hour. The parts count method is used to determine equipment or system failure rate during bid proposal and early design phases when insufficient information is available to use part failure rate estimation models given in MIL-HDBK-217.[4] The method requires the following information

- The generic part types and their quantities
- Equipment/system use environment
- Part quality levels

Under the single use environment, the equipment/system failure rate is expressed by:[4]

$$\lambda_T = \sum_{i=1}^{m} k_i \left(\lambda_g \alpha_g\right)_i , \quad (3.6)$$

where

λ_T is the equipment /system failure rate.
m is the number of different generic part classifications in the equipment/system.
k_i is the generic part quantity.
λ_g is the generic part failure rate expressed in failures/10^6 hours.
α_g is the generic part's quality factor.

The values of λ_g and α_g are tabulated in reference 4.

3.5 MARKOV METHOD

This is a widely used method to evaluate reliability of engineering systems. The method is particularly useful to handle repairable systems and systems with dependent failure and repair modes. The method is subject to the following three assumptions.[8]

- The probability of transition from one system state to another in the finite time interval Δt is $\alpha \Delta t$, where α is the transition rate (i.e., failure or repair rate) from one state to another.
- All occurrences are independent of one another.
- The transitional probability of two or more occurrences in time interval Δt from one system state to another is negligible (e.g., $(\alpha \Delta t)(\alpha \Delta t) \rightarrow 0$).

The application of the method is demonstrated through the following example:

Example 3.3

An X-ray machine was studied for a long period of time, and it was found that its times to failure and repair are exponentially distributed with mean $1/\lambda_x$ and $1/\mu_x$, respectively. More specifically, λ_x and μ_x are constant failure rate and repair rate of the X-ray machine, respectively. The X-ray machine state space diagram is shown in Figure 3.8. The numerals in the boxes denote the X-ray machine state, i.e., operating and failed. Develop expressions for probabilities that the X-ray machine is in up-state and in down-state at time *t* with the aid of the Markov method.

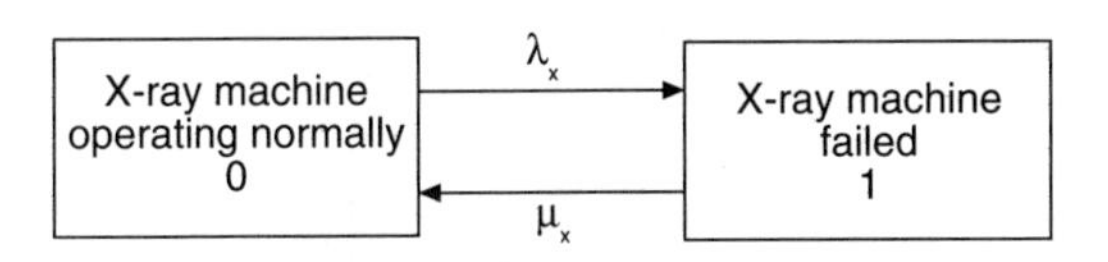

FIGURE 3.8 X-ray machine state space diagram.

With the aid of the Markov method, we write the following equations for Figure 3.8:

$$P_0(t + \Delta t) = P_0(t)\,(1 - \lambda_x \Delta t) + P_1(t)\,\mu_x \Delta t \tag{3.7}$$

$$P_1(t + \Delta t) = P_1(t)\,(1 - \mu_x \Delta t) + P_0(t)\,\lambda_x \Delta t, \tag{3.8}$$

where

$P_i(t)$	is the probability that X-ray machine is in state *i* at time *t*, for i = 0,1.
λ_x	is the X-ray machine failure rate.
μ_x	is the X-ray machine repair rate.
$\lambda_x\, \Delta t$	is the probability of X-ray machine failure in time interval Δt.
$\mu_x\, \Delta t$	is the probability of X-ray machine repair in time interval Δt.
$(1 - \mu_x\, \Delta t)$	is the probability of no repair in time interval Δt.
$(1 - \lambda_x\, \Delta t)$	is the probability of no failure in time interval Δt.
$P_i(t + \Delta t)$	is the probability that X-ray machine is in state *i* at time (t + Δt); for i = 0,1.

Rearranging Equations (3.7) and (3.8), we get

$$\frac{P_0(t+\Delta t) - P_0(t)}{\Delta t} = -P_0(t)\lambda_x + P_0(t)\lambda_x + P_1(t)\mu_x \tag{3.9}$$

$$\frac{P_1(t+\Delta t) - P_1(t)}{\Delta t} = -P_1(t)\mu_x + P_0(t)\lambda_x \tag{3.10}$$

Taking the limit of Equations (3.9) and (3.10) as $\Delta t \rightarrow 0$ results in

$$\frac{dP_0(t)}{dt} = -P_0(t)\lambda_x + P_1(t)\mu_x \tag{3.11}$$

$$\frac{dP_1(t)}{dt} = -P_1(t)\mu_x + P_0(t)\lambda_x \tag{3.12}$$

At time t = 0, $P_0(0) = 1$, $P_1(0) = 0$.

Solving Equations (3.11) and (3.12), we obtain

$$P_0(t) = \frac{\mu_x}{(\lambda_x + \mu_x)} + \frac{\lambda_x}{(\lambda_x + \mu_x)} e^{-(\lambda_x + \mu_x)t} \tag{3.13}$$

$$P_1(t) = \frac{\lambda_x}{(\lambda_x + \mu_x)} - \frac{\lambda_x}{(\lambda_x + \mu_x)} e^{-(\lambda_x + \mu_x)t} \tag{3.14}$$

Equations (3.13) and (3.14) are expressions for probability that the X-ray machine is in up-state and in down-state at time t, respectively.

Example 3.4

Using equations (3.13) and (3.14), obtain expressions for the X-ray machine steady state availability and unavailability. Calculate the X-ray machine availability for a service for a ten-hour mission if its failure and repair rates are 0.005 failures/hour and 0.02 repairs/hour, respectively.

The time-dependent availability and unavailability of the X-ray machine are given by Equations (3.13) and (3.14), respectively. Thus, the X-ray machine steady state availability and unavailability are

$$A_x = \lim_{t \to \infty} P_0(t) = \frac{\mu_x}{\mu_x + \lambda_x} \tag{3.15}$$

and

$$UA_x = \lim_{t \to \infty} P_1(t) = \frac{\lambda_x}{\mu_x + \lambda_x}, \tag{3.16}$$

where

- A_x is the X-ray machine steady state availability.
- UA_x is the X-ray machine steady state unavailability.

Using the given data in Equation (3.13), the X-ray machine availability is

$$P_0(10) = \frac{0.02}{(0.005+0.02)} + \frac{0.005}{(0.005+0.02)} e^{-(0.005+0.02)(10)}$$

$$= 0.9558$$

It means there is approximately a 96% chance that the X-ray machine will be available for service during the ten-hour mission.

3.6 COMMON-CAUSE FAILURE ANALYSIS METHOD

A common-cause failure may be described as any instance where multiple units fail due to a single cause. Thus, the occurrence of common-cause failures leads to lower reliability. Therefore, their consideration in reliability analysis of engineering systems is important. The beginning of the serious thinking on the problem of common-cause failures may be regarded as the late 1960s and the early 1970s when many publications on the subject appeared.[14-18] Some of the causes for the occurrence of common-cause failures in engineering systems are operations and maintenance errors, poor design, common external environments, external catastrophic events, common manufacturer of parts or items, functional deficiencies, and common external power source. References 19–20 present a list of publications on common-cause failures.

Although we have no reliable data on the occurrence of common cause failures specifically in medical devices/equipment, the possibilities of their occurrence are real. The following method will be useful to perform common-cause failure analysis of medical devices/systems:

This is probably the easiest way to analyze redundant systems with common-cause failures. The method is based on the assumption that all common-cause failures associated with a redundant system can be represented by a single hypothetical unit, and in turn, the unit is placed in series with a redundant system as shown in Figure 3.9. The redundant system in the figure is assumed to be a parallel network. Symbols k and cc denote the number of units in parallel and the hypothetical unit representing all system common-cause failures.

The reliability R_n, of the total network shown in Figure 3.9 is given by

$$R_n = [1 - (1 - R)^k]R_{cc} \tag{3.17}$$

where

- R_n is the reliability of the parallel system with common-cause failures.
- k is the number of independent and identical units in parallel.
- R_{cc} is the reliability of the hypothetical unit representing all system common-cause failures.
- R is the unit reliability.

For exponentially distributed unit and common-cause times to failure, a unit total failure rate, λ_T is expressed by

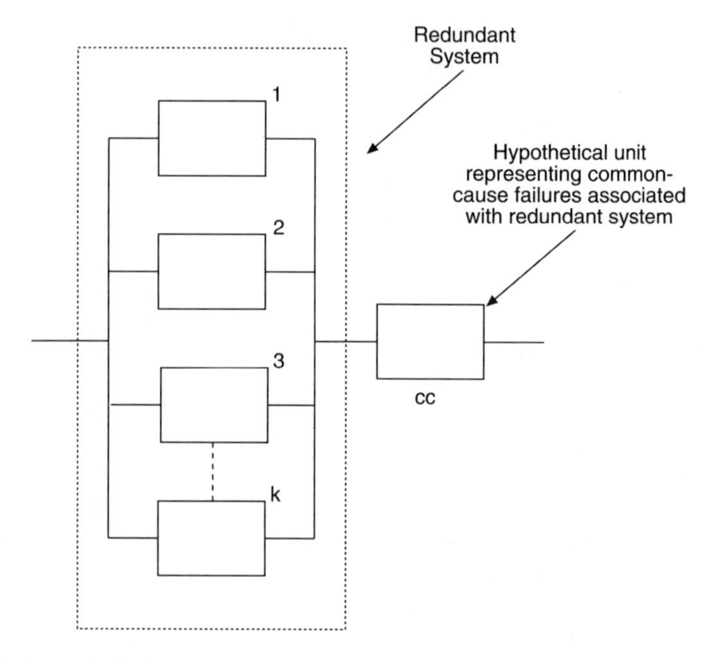

FIGURE 3.9 Block diagram of a redundant system with common-cause failures.

$$\lambda_T = \lambda + \lambda_{cc} \tag{3.18}$$

where

λ is the unit non-common cause failure rate.

λ_{cc} is the hypothetical unit representing all common cause failure rate.

A fraction of unit failures that are common cause is defined by

$$\theta = \frac{\lambda_{cc}}{\lambda_T} \tag{3.19}$$

Therefore,

$$\lambda_{cc} = \theta\, \lambda_T. \tag{3.20}$$

Inserting Equation (3.20) into Equation (3.18) yields

$$\lambda = (1 - \theta)\, \lambda_T \tag{3.21}$$

Thus, with the aid of Equations (3.20) and (3.21), we get

$$R_{cc}(t) = e^{-\theta \lambda_T t} \tag{3.22}$$

and

$$R(t) = e^{-(1-\theta)\lambda_T t}, \qquad (3.23)$$

where

t is time.

Using Equations (3.22) and (3.23) in Equation (3.17) yields

$$R_n(t) = \left[1 - \left(1 - e^{-(1-\theta)\lambda_T t}\right)^k\right] e^{-\theta\lambda_T t} \qquad (3.24)$$

By integrating Equation (3.24) over the interval [0, ∞], we get the following expression for the system mean time to failure:

$$MTTF_n = \sum_{i=1}^{k} (-1)^{i+k} \binom{k}{i} \Big/ \lambda_T \left[i - (i-1)\theta\right], \qquad (3.25)$$

where

$MTTF_n$ is the mean time to failure of the parallel system with common cause failures.

$$\binom{k}{i} = \frac{k!}{i!(k-i)!} \qquad (3.26)$$

Example 3.5

Assume that a piece of medical equipment contained two independent and identical subsystems in parallel. The parallel system can also fail due to the occurrence of common-cause failures. The noncommon-cause failure rate of a subsystem is 0.001 failures/hour, and the common-cause failure rate of the parallel system is 0.0005 failures/hour. Calculate the medical equipment parallel system reliability for a 100-hour mission with and without the occurrence of common-cause failures.

Using the given data in Equations (3.18) and (3.19), we get

$$\lambda_T = 0.001 + 0.0005$$

$$= 0.0015 \text{ failures/hour}$$

and

$$\theta = 0.0005/0.0015$$

$$= 0.3333$$

Substituting the above calculations and the given data into Equation (3.24) yields

$$R_n(100) = \left[1 - \left(1 - e^{-(1-0.3333)(0.0015)(100)}\right)^2\right] e^{-(0.3333)(0.0015)(100)}$$

$$= 0.9426$$

When there are no common-cause failures, $\lambda_{cc} = 0$ and $\theta = 0$. And from Equations (3.18) and (3.24), we get

$$R_n(t) = 1 - (1 - e^{-\lambda t})^k \tag{3.27}$$

Using the given data in Equation (3.27) yields

$$R_n(100) = 1 - \left(1 - e^{-(0.001)(100)}\right)^2$$

$$= 0.9909$$

It means the medical equipment parallel system reliability with and without common-cause failures is 0.9426 and 0.9909, respectively.

Example 3.6

In example 3.5, calculate the medical equipment parallel system mean time to failure with and without the occurrence of common cause failures. Comment on the results. By substituting the given data into Equation (3.25), we get

$$MTTF_n = \frac{2}{\lambda_T} - \frac{1}{\lambda_T(2-\theta)}$$

$$= (2/0.0015) - (1/(.0015)(2 - 0.3333))$$

$$= 933.34 \text{ hours}$$

For no common-cause failures, i.e., $\lambda_{cc} = \theta = 0$ and other given data, Equation (3.25) simplifies to

$$MTTF_n = \frac{2}{\lambda} - \frac{1}{2\lambda}$$

$$= (2/0.001) - (1/2(0.001))$$

$$= 1500 \text{ hours}$$

It means the mean time to failure of the medical equipment parallel system with common-cause failures is 933.34 hours and without common cause failures 1500 hours. Obviously, the occurrence of common-cause failures substantially reduced the system mean time to failure, i.e., from 1500 hours to 933.34 hours.

3.7 PROBLEMS

1. What is the difference between FMEA and FMECA?
2. Outline the steps associated with the performance of FMEA.
3. What are the advantages of FMEA?
4. Define the following terms:
   - Single failure point
   - Failure mode
   - Common-cause failure
   - AND gate
5. Describe the steps associated with performing FTA.
6. What are the benefits and drawbacks of FTA?
7. What are the causes for the occurrence of common-cause failures?
8. Assume that the failure and repair rates of an ultrasound device are 0.0002 failures/hour and 0.004 repairs/hour, respectively. Calculate the device steady state availability and unavailability.
9. Prove that the sum of Equations (3.13) and (3.14) is equal to unity, i.e., $P_0(t) + P_1(t) = 1$.
10. Compare FMEA and FTA methods.

REFERENCES

1. MIL-F-18372(Aer.), General Specification for Design, Installation, and Test of Aircraft Flight Control Systems, Bureau of Naval Weapons, Department of the Navy, Washington, DC, para. 3.5.2.3.
2. Coutinho, J.S., Failure Effect Analysis, *Trans. NY Acad. Sci.*, Vol. 26, Series II, 1963-64, pp. 564–584.
3. Haasl, D.F., Advanced Concepts in Fault Tree Analysis, System Safety Symposium, 1965. Available from the University of Washington Library, Seattle.
4. MIL-HDBK-217, Reliability Prediction of Electronic Equipment, Department of Defense, Washington, DC.
5. Fleming, K.N., A Redundant Model for Common Mode Failures in Redundant and Safety Systems, *Proceedings of the Sixth Pittsburgh Annual Modeling and Simulation Conference*, 1975, pp. 579–581.
6. Dhillon, B.S. and Proctor, C.L., Common Mode Failure Analysis of Reliability Networks, *Proceedings of the Annual Reliability and Maintainability Symposium*, 1977, pp. 404–408.
7. MIL-STD-1629, Procedures for performing a Failure Mode, Effects, and Criticality Analysis, Department of Defense, Washington, DC, 1980.

8. Grant Ireson, W., Coombs, C.F., and Moss, R.Y., *Handbook of Reliability Engineering and Management*, McGraw-Hill Book Company, New York, 1996.
9. Dhillon, B.S., *Systems Reliability, Maintainability and Management*, Petrocelli Books, New York, 1983.
10. Dhillon, B.S., Failure Mode and Effects Analysis: Bibliography, *Microelectronics and Reliability*, Vol. 32, 1992, pp. 719–731.
11. Fussell, J.E., Powers, G.J., and Bennets, R.G., Fault Trees: A state of the Art Discussion, *IEEE Transactions on Reliability*, Vol. 30, 1974, pp. 51–55.
12. Schroder, R.J., Fault Tree for Reliability Analysis, *Proceedings of the Annual Symposium on Reliability*, 1970, pp. 200–206.
13. RDH-376, Reliability Design Handbook, Reliability Analysis Center, Rome Air Development Center, Griffis Air Force Base, Rome, NY, 1976.
14. Ditto, S.J., Failures of Systems Designed for High Reliability, *Nuclear Safety*, Vol. 8, 1966, pp. 35–37.
15. Elper, E. P., Common-Mode Considerations in the Design of Systems for Protection and Control, *Nuclear Safety*, Vol. 11, 1969, pp. 323–327.
16. Gangloff, W.C., Common-Mode Failure Analysis is "In", *Electronic World*, October 1972, pp. 30–33.
17. Elper, E.P., The ORR Emergency Cooling Failures, *Nuclear Safety*, Vol. 11, 1970, pp. 323–327.
18. Jacobs, I.M., The Common-Mode Failure Study Discipline, *IEEE Transactions on Nuclear Science*, Vol. 17, 1970, pp. 594–598.
19. Dhillon, B.S., On Common-Cause Failures: Bibliography, *Microelectronics and Reliability*, Vol. 18, 1978, pp. 533–535.
20. Dhillon, B.S. and Anude, O.C., Common Cause Failures in Engineering Systems: A Review, *Int. J. Reliability, Quality, Safety Eng.*, Vol. 1, 1994, pp. 103–129.

4 Human Error in Health Care Systems

CONTENTS

4.1 INTRODUCTION

Human errors are universal and committed each day. Usually, most are trivial, but some can be very serious or fatal. Two well-publicized examples of disasters due to the human error are the Three-Mile-Island and Chernobyl nuclear accidents. Even though the precise data on the extent of human error as a percentage of all system failures are difficult to obtain and vary from one system to another, one study indicates that it can vary from 90% in air traffic control to 19% in the petroleum industry.[1] The figures for the other sectors of the industry were automobiles (85%),

U.S. nuclear power plants (70%), worldwide jet cargo transport (65%), and petrochemical plants (31%). Nonetheless, the average across all systems is approximately 60%.

In all technical medical equipment problems, operator errors account for well over 50%.[2] Furthermore, as per references 3–5, human error is considered to contribute or cause up to 90% of accidents both generally and in medical devices. In 1984, it was estimated that the cost of the consequences of preventable adverse events in hospitalized patients in the U.S. was in the order of $24.5 billion.[6]

This chapter presents different aspects of human error in health care systems.

4.2 FACTS, FIGURES, AND EXAMPLES OF HUMAN ERROR IN HEALTH CARE SYSTEMS

Over the years, the occurrence of various types of human errors in health care has been recorded and many studies have been conducted. Some of the facts, figures, and examples concerning these human errors are

- A study of anesthesiology errors revealed that 23% of the anesthesiologists said they committed an error with fatal results.[7]
- In 1992, it was estimated that avoidable mistakes kill 100,000 patients a year in the U.S.[8]
- The result of a study indicates that avoidable deaths from anesthesia-related incidents range from 2000 to 10,000 per year in the U.S.[9-11]
- Several studies indicate that human error was a factor in 65% to 87% of deaths attributable to anesthesia.[12 15]
- A study of one of the largest medical liability insurance carriers revealed that one or two malpractice claims are filed for every 10,000 anesthetics administered.[16]
- A survey of anesthetic incidents in the operating room attributed between 70% and 82% of the incidents to the human element.[17-18]
- The results of several studies indicates that human errors such as "inadvertent gas flow change" or "syringe swap" could account for up to 70% of anesthetic mishaps.[15,16,19]
- A study performed in 1985 indicates that 75% of the intra-operative cardiac arrests investigated were preventable.[16,20]
- A study of 5612 surgical admissions to a hospital revealed 36 were identified as adverse outcomes due to human error.[21]
- Deaths or serious injuries associated with medical devices reported through the Food and Drug Administration's (FDA's) Center for Devices and Radiological Health (CDRH) accounted approximately 60% to user error.[22]
- An investigation of 2.7 million patients discharged from hospitals in New York in 1984 revealed 98,609 suffered an adverse event. A further study showed that 25% of these adverse events were due to negligence.[23]
- Another study revealed that more than 35% of medical malpractice claims closed in 1984 were generated in the operating room, and the claims

represented more than 55% of the entire cost of medical malpractice indemnity payments in that year. This percentage in dollars was $1.4 billion, thus averaging more than $110,000 per claim.[24]

- A group of medical device manufacturers conducted an investigation of design validation testing errors and found 40% of the errors were caused by coding or schematics implementation, 40% by changes in requirements, 20% by incorrect test protocols or misinterpretation by the tester, and none by design.[25]
- In 1990, a heart patient in a hospital in New York City was incorrectly given 120 cubic centimeters a minute of a powerful drug instead of 12 cubic centimeters an hour. The patient died.[26]
- A human error caused a fatal radiation-overdose accident involving the Therac radiation therapy device.[27]
- During the treatment of an infant patient with oxygen, a physician set the flow control knob between 1 and 2 liters per minute without realizing the fact that the scale numbers represented discrete, instead of continuous, settings. As the result of the physician's action, the patient became hypoxic.[28]
- A patient was seriously injured by over-infusion because a nurse mistakenly read the number 7 as 1.[28]
- A study of patients undergoing laparoscopic gall bladder surgery in New York indicates that 2% of the patients were injured due to surgeon errors.[29]
- A Hasidic scholar died because emergency room doctors failed to notice a stab wound in his back.[26]
- A study conducted in 1991 found that in a medication calculation test administered to 110 nurses, 81% failed to calculate medication dosages accurately 90% of the time.[30-31]
- A study performed by the New York State Laboratory regulators discovered testing errors in 66% of the laboratories offering drug-screening services.[1,32]
- Two good examples of errors in radiotherapy happened when a patient was administered 50 millicurie instead of 3 millicurie prescribed by the physician and the wrong patient was administered 100 rad to the brain.[1]
- A hospital has to pay $375,000 as a settlement amount after a patient died due to an infusion pump being incorrectly set to deliver 210 cc of heparin per hour rather than the ordered 21 cc.[33]

4.3 CAUSES OF PATIENT INJURIES AND MEDICAL DEVICE ACCIDENT CLASSIFICATION

Before we discuss the topic of human error further, let us examine the causes of patient injuries and medical device accident classification because both subjects directly or indirectly relate to the occurrence of human error.

Over the years, professionals working in the medical field have identified eight fundamental mechanisms, as shown in Figure 4.1, through which patients (or

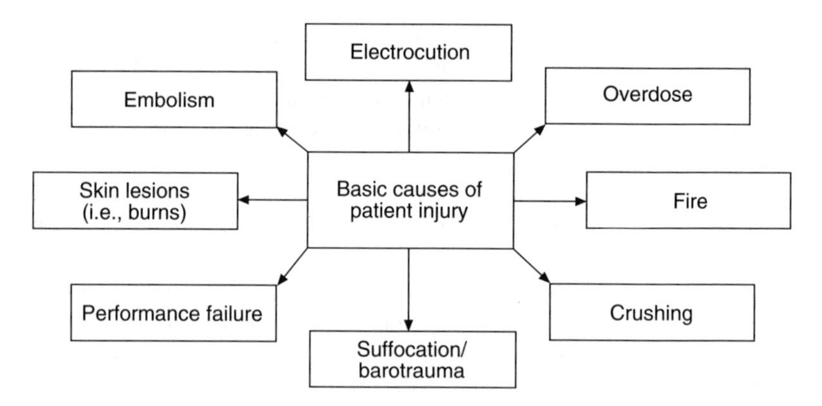

FIGURE 4.1 Basic causes of patient injury.

personnel) may be injured or killed.[33] The understanding of these mechanisms or causes is essential to determine the extent to which a device's design or use may have contributed to the occurrence of an accident.

Medical device accidents may be classified into seven categories: operator or patient error, manufacturing defect, design deficiency, abnormal or idiosyncratic patient response, faulty preventive maintenance, repair or calibration, random component failure, and sabotage or malicious intent.[33]

4.4 MEDICAL DEVICES WITH A HIGH OCCURRENCE OF HUMAN ERROR

Over the years, many studies have been performed to identify medical devices that have a high incidence of human error. As the result of these studies, the most error-prone medical devices were identified.[29] These devices include (in the order of most error-prone to least error-prone) glucose meter, balloon catheter, orthodontic bracket aligner, administration kit for peritoneal dialysis, permanent pacemaker electrode, implantable spinal cord simulator, intra-vascular catheter, infusion pump, urological catheter, electrosurgical cutting and coagulation device, nonpowered suction apparatus, mechanical/hydraulic impotence device, implantable pacemaker, peritoneal dialysate delivery system, catheter introducer, catheter guidewire, transluminal coronary angioplasty catheter, external low-energy defibrillator, continuous ventilator (respirator), and contact lens cleaning and disinfecting solutions. According to reference 29, the errors in using medical devices cause, on average, at least three deaths or serious injuries each day.

4.5 ANESTHESIA-RELATED MOST FREQUENT INCIDENTS, ASSOCIATED FACTORS, AND MEDICAL DEVICE-RELATED OPERATOR ERRORS

As past studies indicate, anesthesia-related mishaps are a major problem. Reference 15 developed a list of ten most frequent incidents. Thus, in the order of the frequency of their occurrence (from highest to lowest) anesthesia-related mishaps include: breathing circuit disconnection, inadvertent gas flow change, syringe swap, gas supply problem, intravenous apparatus disconnection, laryngoscope malfunction, premature extubation, breathing circuit connection error, hypovolemia, and tracheal airway device position changes.

In the order of the frequency of their occurrence (from highest to lowest) associated factors are inadequate total experience, inadequate familiarity with equipment/device, poor communication with team, lab, etc., haste, inattention/carelessness, fatigue, excessive dependency on other personnel, failure to perform a normal check, training or experience, supervisor not present enough, environment or colleagues, visual field restricted, mental or physical, inadequate familiarity with anaesthetic technique, teaching activity under way, apprehension, emergency case, demanding or difficult case, boredom, nature of activity, insufficient preparation, and slow procedure.[15]

There are numerous operator-related errors that occur during the operation or maintenance of medical device/equipment. The most important are listed in Table 4.1.[34]

TABLE 4.1
Important Operator Errors Associated With Medical Devices/Equipment

No.	Error Description
1	Errors in setting device parameters.
2	Departure from following stated instructions and procedures.
3	Incorrect selection of devices with respect to the clinical objectives and requirements.
4	Untimely/inappropriate/inadvertent activation of controls.
5	Inappropriate improvisation.
6	Misassembly.
7	Misinterpretation of or failure to recognize vital device outputs.
8	Over-reliance on a medical device's/equipment's capabilities, automatic features, or alarms.
9	Wrong decision-making and actions to critical situations.

4.6 GENERAL APPROACH TO HUMAN FACTORS IN THE MEDICAL DEVICE DEVELOPMENT PROCESS TO REDUCE HUMAN ERRORS

The occurrence of human errors can be significantly reduced by making human factors an integral part of the medical device development process from the concept phase to the production phase as shown in Figure 4.2.[35]

During the concept phase, the human factors specialist works with market researchers, helps develop and implement questionnaires, conducts interviews with potential users of the device, evaluates competitive devices, and performs analysis of industry and regulatory standards. The specialist also examines the proposed operation of the potential device with respect to educational background, skill range,

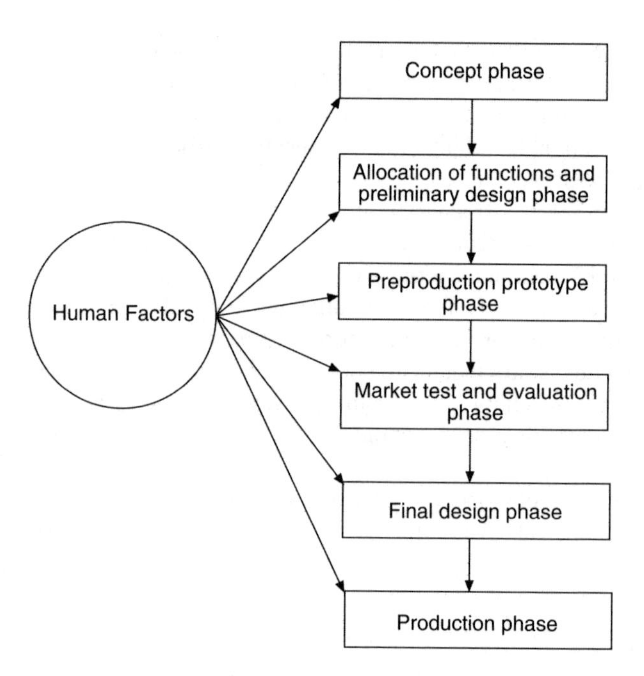

FIGURE 4.2 Human factors inputs during the medical device development process.

and experiences of the intended users and identifies the possible use environments of the device under consideration.

During the allocation of functions and preliminary design phase, both the human factors specialist and the design professionals determine which device functions will be automatic and which will require manual points of interface between humans and the device. More specifically, the points of interface are those operations where humans have to monitor and control so that the desired output or feedback from the device is obtained. The analysis of the preliminary design is performed with respect to the device operating environment and the skill level of the most untrained user. Usually, this task is performed by considering the drawings or sketches of the operational environment and gauging reactions of potential users.

After evaluating the device, during the preproduction prototype phase, the prototype is built or updated for additional evaluation and market testing.

The market test and evaluation phase involves not only the actual testing of the device, but also a thorough examination of the feedback received from the market test by human factors, marketing, and engineering professionals.

During the final design phase, the device design is finalized by incorporating any human-factors-related changes generated by the marketing, test, and evaluation.

In the production phase, the device is produced and put on the market. Nonetheless, during this phase the human factors specialist usually monitors the device performance, conducts analysis of the proposed design changes, and assists in the development of user-related training programs.

4.7 HUMAN ERROR-RELATED ANALYSIS METHODS USEFUL FOR HEALTH CARE SYSTEMS

Over the years, many techniques and methods have been developed to perform human error/reliability related analysis for the purpose of reducing or completely eliminating the occurrence of human error in engineering systems. These methods can be applied equally to health care systems or devices. Four of the methods are described below.

4.7.1 FAILURE MODE AND EFFECT ANALYSIS (FMEA)

This is one of the most widely used methods to analyze engineering systems during their design, and it may simply be described as a structured analysis of an item or function that highlights potential failure modes, their causes, and the effects of a failure on system operation.

When FMEA also evaluates the criticality of the failures — that is, the severity of the effect of the failure and the probability of its occurrence — the analysis is referred to as failure mode, effects, and criticality analysis (FMECA) and the failure modes are assigned priorities.

The FMEA technique was developed in the early 1950s to analyze flight control systems.[36] During the course of the next decade, FMECA grew out of FMEA, and in the 1970s, the U.S. Department of Defense developed a military standard[37] entitled "Procedures for Performing a Failure Mode, Effects, and Criticality Analysis". The most important revision of the document occurred in 1984.[38] A comprehensive list of publications on FMEA and FMECA is given in reference 39.

There are seven main steps involved in performing FMEA:

- Establish system definition
- Establish ground rules
- Describe system hardware
- Describe functional blocks
- Identify failure modes and their effects
- Compile critical items list
- Document

The method is described in detail in Chapter 3.

4.7.2 THROUGHPUT RATIO METHOD

This method was developed by the Navy Electronics Laboratory Center. It determines the operability of man-machine interfaces or stations.[40] A typical example of the interfaces or stations is the control panel. The operability may be described as the extent to which the man-machine station performance meets the design expectation for the station under consideration.[41]

The term "throughput" means transmission because the ratio is expressed in terms of responses or items per unit time emitted by the operator. Thus, the throughput ratio is defined as follows:

$$TR = [(m/n) - CF]\,(100) \tag{4.1}$$

where

- TR is the man-machine operability expressed in percentage.
- m is the number of throughput items generated per unit time.
- n is the number of throughput items to be generated per unit time to meet design expectation.
- CF is the correction factor. More specifically, correction for error or out-of-tolerance output.

In turn, CF, is given by

$$CF = D_1 D_2 \tag{4.2}$$

where

$$D_1 \equiv (T_i/T)\,(m/n) \tag{4.3}$$

$$D_2 = D_1 P_n^2 P_f \tag{4.4}$$

The symbols used in Equations (4.3) and (4.4) are described below.

- T_i is the number of trials in which the control-display operation is performed incorrectly.
- T is the total number of trials in which the control-display operation is performed.
- P_n is the probability that the operator will not detect the error.
- P_f is the probability of function failure because of an error.

Some of the applications of the throughput ratio method can be used to determine system acceptability, redesign of the evaluated design with respect to human factors, establish system feasibility, and perform comparisons of alternative design operabilities.

Example 4.1

For the following given data, calculate the value of the man-machine operability ratio.

$$m = 5,\ n = 10,\ P_n = 0.5,\ P_f = 0.7,\ T_i = 4, \text{ and } T = 16$$

By inserting the above data into Equations (4.2)–(4.4), we get

$$CF = D_1 D_2 = (0.125)(0.0219) = 0.0027$$

$$D_1 = (4/16)(5/10) = 0.125$$

$$D_2 = (0.125)(0.5)^2 (0.7)$$

$$= 0.0219$$

Substituting the above result and the other given data into Equation (4.1) yields

$$TR = [(5/10) - 0.0027](100)$$

$$= 49.73\%$$

Thus, the value of the man-machine operability ratio is 49.73%.

4.7.3 Probability Tree Method

This method is useful to represent critical human actions and other events associated with the system under study. The branches of the probability tree represent the diagrammatic task analysis. In addition, the branching limbs of the tree represent each event's outcome, i.e., success or failure. The probability of occurrence is assigned to each tree branch. Some advantages of the method are as follows:[42]

- A visibility tool.
- The probability of occurrence of errors due to computation decreases because of simplification in mathematical computations.
- Factors such as interaction of stress, emotional stress, and interaction effects can be incorporated with some modifications.
- The analyst can readily estimate conditional probability which may otherwise be estimated through rather complicated probability equations.

The method is demonstrated through Example 4.2.

Example 4.2

Assume a medical equipment operator performs two independent tasks, say, x_1 and x_2. Both tasks can either be performed correctly or incorrectly. Thus, in this case, the incorrectly performed tasks are the only errors that may occur. Obtain an expression for the overall system probability of performing incorrect task by developing a probability/event tree. Assume that task x_1 is performed before task x_2.

The probability/event tree for the example is shown in Figure 4.3 where x_1 and x_2 denote success events of performing tasks x_1 and x_2, respectively, and $\bar{x}_1$ and $\bar{x}_2$ - failure events of tasks x_1 and x_2, respectively. With the aid of Figure 4.3, the overall system probability of performing incorrect tasks is

$$P_{inc} = P_{x_1} P_{\bar{x}_2} + P_{\bar{x}_1} P_{x_2} + P_{\bar{x}_1} P_{\bar{x}_2} \tag{4.5}$$

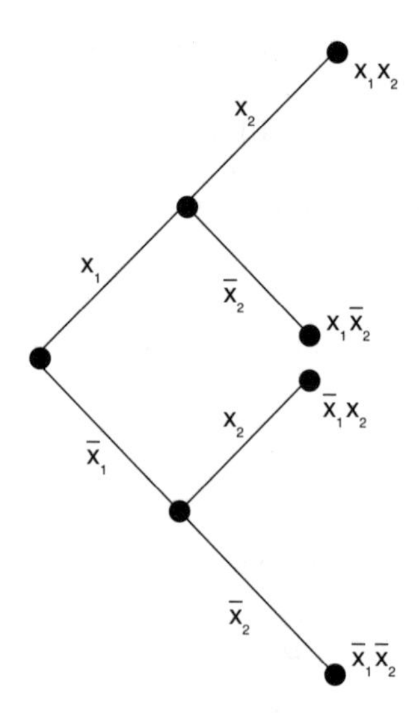

FIGURE 4.3 Probability/event tree.

where

P_{inc} is the overall system probability of performing incorrect task (i.e, performing both tasks x_1 and x_2 incorrectly).

P_{x_i} is the probability of performing task x_i correctly; for i = 1, 2.

$P_{\bar{x}_i}$ is the probability of performing task x_i incorrectly; for i = 1, 2.

Since $P_{x_1} + P_{\bar{x}_1} = 1$, and $P_{x_2} + P_{\bar{x}_2} = 1$, Equation (4.5) simplifies to

$$\begin{aligned} P_{inc} &= \left(1 - P_{\bar{x}_1}\right) P_{\bar{x}_2} + P_{\bar{x}_1}\left(1 - P_{\bar{x}_2}\right) + P_{\bar{x}_1} P_{\bar{x}_2} \\ &= 1 - \left(1 - P_{\bar{x}_1}\right)\left(1 - P_{\bar{x}_2}\right) \end{aligned} \tag{4.6}$$

4.7.4 Fault Tree Method

This a widely used method in the industry to perform reliability analysis of engineering systems. The method was developed in the early 1960s at Bell Laboratories to perform failure analysis of the Minuteman Launch Control System. It can also be applied to perform human error analysis of medical devices.

Fault tree analysis starts by identifying an undesirable event, called top event, associated with a system. Events that could cause the top event are generated and connected by logic operators such as AND and OR. The AND gate provides a true (failed) output if all the inputs are true (failures). The OR gate provides a true (failure) output if one or more inputs are true (failures). The fault tree construction proceeds by generation of events in a successive manner until the events (basic events) do not need to be developed further. The fault tree is the logic structure relating the top event to the basic events.

The basic steps involved in performing fault tree analysis are as follows:

- Define the system, the assumptions involved in the analysis, and the events or states that would constitute failure.
- Establish a system block diagram indicating inputs, outputs, and interfaces if simplification of the scope of analysis is desirable.
- Establish the top-level fault event.
- Use fault tree logic and fault event symbols and apply deductive reasoning to identify what could cause the top-level fault event to occur.
- Continue developing the logic tree by identifying causes for intermediate fault events, that is, the fault events that can cause the top-level fault event.
- Develop the fault tree to the desired lowest level, that of the most basic fault events.
- Analyze the completed fault tree qualitatively and quantitatively.
- Identify appropriate corrective measures.
- Document the analysis and take appropriate measures to rectify problem areas.

This method is described in detail in Chapter 3.

Example 4.3

Assume that a medical device operator is required to perform a task, say Z, which is composed of two independent subtasks, y_1 and y_2. Both subtasks must be performed correctly for the success of the task. In turn, subtasks, y_1 is composed of two steps, x_1 and x_2, and each step must be performed correctly for the successful performance of subtask y_1.

Similarly, subtask y_2 is made up of two steps, S_1 and S_2, and if at least one of these steps is performed correctly, subtask y_2 will be accomplished successfully. Construct a fault tree with the top event entitled "The operator will perform task Z incorrectly" if all the steps involved are independent. Also, calculate the probability of the occurrence of the top event if the unsuccessful occurrence probability of each step is 0.1.

Figure 4.4 presents fault tree for the top event "The operator will perform task Z incorrectly." In the same figure, the probability of occurrence of the identified events is shown in parentheses.

The unsuccessful probability of subtask y_1 performance is given by

$$P(y_1) = P(x_1 + x_2) = P(x_1) + P(x_2) - P(x_1)P(x_2)$$

$$= (0.1) + (0.1) - (0.1)(0.1)$$

$$= 0.19$$

Similarly, the unsuccessful probability of subtask y_2 performance is

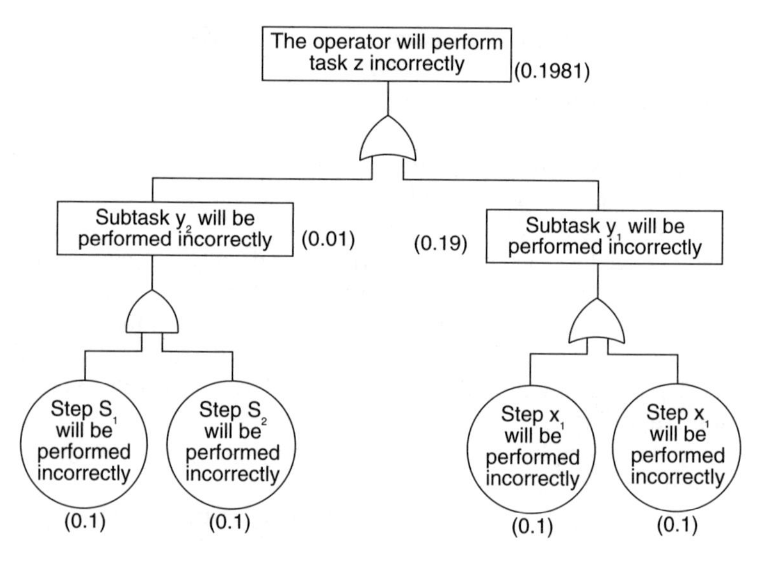

FIGURE 4.4 Fault tree for the medical device operator performing task Z incorrectly.

$$\begin{aligned} P(y_2) &= P(S_1)P(S_2) \\ &= (0.1)(0.1) \\ &= 0.01 \end{aligned}$$

The probability that the operator will perform task Z incorrectly is given by

$$\begin{aligned} P(Z) = P(y_1 + y_2) &= P(y_1) + P(y_2) - P(y_1)P(y_2) \\ &= (0.19) + (0.01) - (0.19)(0.01) \\ &= 0.1981 \end{aligned}$$

This means that there is approximately a 20% chance that the medical device operator will perform task Z incorrectly.

4.8 REDUCING MEDICAL DEVICE/EQUIPMENT USER INTERFACE-RELATED ERRORS

This section presents two sets (i.e., one comprehensive and the other general) of guidelines to reduce medical device user interface errors. The comprehensive set of guidelines are basically rules of thumb developed by the FDA's CDRH to reduce the occurrence of human errors related to areas such as control/display arrangement and design, software design, component installation, and alarms.[28] The general set of guidelines is the ten design tips to enhance user interfaces and improve medical device usability.

4.8.1 Comprehensive Guidelines

Control/Display Arrangement and Design

An example of a human error in this area is that a physician treating an infant patient with oxygen set the flow control knob between 1 and 2 liters per minute without realizing that the scale numbers represented discrete instead of continuous, settings.[28] The following rules of thumb for designing the user interface can help reduce the occurrence of human error:

- Design control knobs and switches so that they correspond to user conventions.
- Ensure that all facets of design are compatible with user expectations as much as possible.
- Make the intensity and pitch of auditory signals easily heard by the device users.
- Ensure that tactile feedback is provided by controls.
- Ensure that the occurrence of inadvertent activation is reduced to its minimal level with respect to the arrangement and design of knobs, switches, and keys.
- Ensure that controls, workstations, and display are designed around the basic capabilities of users, i.e., strength, vision, hearing, memory, reach, etc.
- Design displays and labels so that they can easily be read from typical distances and angles.
- Ensure that control and display arrangements are well-organized and uncluttered.
- Ensure that keys, switches, and control knobs are spaced apart sufficiently for easy manipulation.
- Make use of color and shapes coding, as appropriate, to facilitate the rapid identification of controls and displays.
- Ensure that symbols, test, abbreviations, and acronyms placed on or displayed by the device are compatible with the ones in the instructional manual.
- Design the brightness of visual signals so that it is adequate for users performing their tasks under varying conditions of ambient illumination.

Software Design

An example of human error related to the software design is that a cardiac output monitor alarm was turned off without the operator's knowledge when the control buttons were pushed in a certain order.[28] Some of the rules of thumb that can help prevent software design-related human errors, are as follows:

- Avoid contradicting expectations of users.
- Ensure that the design of headings, symbols, and formats are consistent.
- Develop procedures that entail easy-to-remember steps.

- Avoid overloading users with unformatted, densely packed, or insufficient information.
- Avoid using software in situations where a simple hardware solution is feasible.
- Provide users recourse in the event of an error.
- Ensure that standard symbols, colors, icons, and abbreviations are used to convey information reliably and efficiently.
- Ensure that dedicated displays or display sectors are used for highly critical information.
- Keep users up-to-date concerning the current device status.
- Ensure that prompt and clear feedback is provided following user entries.

Component Installation

An example of human error related to the component installation is that a part of an oxygen machine was installed upside down, which resulted in the death of a patient due to the impediment of air flow.[28] The following are guidelines for reducing the likelihood of confusion between similar items or parts and making wrong connections:

- Make instructions to users simple and straightforward and warnings conspicuous.
- Design connectors, cables, tubing, and other hardware for easy installation and connection.
- Add graphics to reduce textural complexity in maintenance manuals.
- Design for positive locking mechanisms if the integrity of connections is expected to be compromised due to factors such as motion, casual contact, or component durability.
- Ensure that components and accessories are numbered so that defective parts can easily be replaced with good ones.
- Ensure that in the event it is impossible to eliminate a hazard through design, color codes or other markings, the user achieves proper connections and component or accessory installation.
- Protect body lead electrical contacts, and, if possible avoid exposed contacts altogether.

Alarms

An example of a human error associated with alarms happened when a patient receiving oxygen died when a concentrator pressure hose became loose and the alarm was not loud enough to be heard over the drone of the device.[28] Some rules of thumb that reduce the occurrence of human errors associated with alarms are

- Ensure that alarms effectively satisfy normal hearing and visual limits of the average user.
- Consider a wide spectrum of operational environments during alarm designing and testing.

- Ensure that the critical alarms receive a priority status.
- Include visual and auditory alerts and critical alarms in the design requirements for the device.
- Ensure that all used codes correspond to established conventions.
- Examine with care the effects of electromagnetic interference, over-sensitivity, and static electricity on the functioning of alarms.
- Ensure that the alarms in their silence mode remain silent.
- Ensure that all alarms used are distinguishable from one another.
- Ensure that color contrast and brightness contrast are adequate for an effective legibility under varying lighting conditions.
- Ensure that all alarms used can activate immediately after a critical problem is detected.

4.8.2 General Guidelines

These guidelines are concerned with making medical device interfaces more user-friendly because user-interface designs such as those associated with items such as ventilators, infusion pumps, patient monitors, kidney dialysis machine, and blood chemistry analyzers frequently experience superficial design problems that can lead to negative effects on a device's usability and appeal.[43] Nonetheless, past experience shows that problems associated with medical device user-interface designs often lead to the occurrence of human errors.

Ten design tips presented below can help alleviate medical device user-interface design problems.[43]

- **Decrease or reduce screen density**. This involves reducing the over-stuffing of medical device displays with information because it could be difficult for nurses, physicians, etc. to pick out information at a glance from overly dense user interfaces.
- **Use simple and straightforward language**. This means using simple language because medical device user interfaces often have overly complex words and phrases.
- **Establish visual balance**. This involves creating visual balance or symmetry about the vertical axis. More specifically, arrange visual elements on either side of an assumed axis so that each side has approximately the same amount of content as empty space.
- **Simplify typography**. This eliminates excessive highlighting such as italicized, bolded, and underlined words.
- **Provide navigation cues and options**. Moving from one place to another in a medical device user interface can sometimes cause a user to become lost. Thus it is important to provide navigation cues and options.
- **Limit the number of colors**. Past experience indicates that it is useful to limit the color palette of a user interface and keep background and major on-screen components between three and five colors, including shades of grey. Also, it is important to select colors carefully so that they are consistent with medical conventions.

- **Ascribe to a grid**. This involves fitting on-screen elements into a grid because it eventually pays off in terms of visual appeal and perceived simplicity. Also, grid-based screens tend to be easy to implement in computer code because of the predictability of the visual elements' position.
- **Make use of hierarchical labels**. Because redundant labelling usually leads to congested screens and takes a long time to scan, hierarchical labelling can save space and speed scanning by displaying items such as respiratory rate, arterial blood pressure, and heart rate more efficiently.
- **Avoid inconsistencies**.
- **Refine and harmonize icons**.

4.9 PROBLEMS

1. Define the term "human error."
2. Give at least five examples of the occurrence of human error in health care systems.
3. What are the basic causes of patient injuries?
4. What are the typical classifications of medical device accidents?
5. List at least ten medical devices that have a high occurrence of human error.
6. Write down at least five important operator-related errors associated with medical devices.
7. Describe the following human error-related analysis methods useful for the health care systems:
   - The throughput ratio method
   - The probability tree method
8. Give examples of human errors associated with the following areas:
   - Control/display arrangement and design
   - Component installation
   - Software design
   - Alarms
9. List at least ten guidelines to reduce the occurrence of human errors with control/display arrangement and design.
10. Discuss at least five rules of thumb/guidelines to reduce the likelihood of confusion between similar parts and making wrong connections.

REFERENCES

1. Van Cott, H., Human Errors: Their Causes and Reduction, *Human Error in Medicine*, M.S. Bogner, Ed., Lawrence Erlbaum Associates Publishers, Hillsdale, NJ, 1994, pp. 53–65.
2. Dhillon, B.S., Reliability Technology in Health Care Systems, *Proceedings of the IASTED International Symposium on Computers and Advanced Technology in Medicine, Health Care, and Bio-engineering*, 1990, pp. 84–87.

3. Nobel, J.L., Medical Device Failures and Adverse Effects, *Pediat. Emerg. Care*, Vol. 7, 1991, pp. 120–123.
4. Bogner, M.S., Medical Devices and Human Error, *Human Performance in Automated Systems: Current Research and Trends*, M. Mouloua and R. Parasuraman, Eds., Hillsdale, Lawrence Erlbaum Associates Publishers, Hillsdale, NJ, 1994, pp. 64–67.
5. Maddox, M.E., Designing Medical Devices to Minimize Human Error, *Med. Dev. Diag. Ind. Mag.*, Vol. 19, No. 5, 1997, pp. 166–180.
6. Bogner, M.S., Human Error in Medicine: A Frontier for change, Human Error in Medicine, M.S. Bogner, Ed., Lawrence Erlbaum Associates, Publishers, Hillsdale, NJ, 1994, pp. 373–383.
7. McDonald, J.S. and Peterson, S., Lethal Errors in Anaesthesiology, *Anaesthesiology,* Vol. 63, 1985, pp. A 497.
8. Russell, C., Human Error: Avoidable mistakes kill 100,000 patients a year, *The Washington Post*, February 18, 1992, pp. WH 7.
9. Cooper, J.B., Toward Prevention of Anaesthetic Mishaps, *Analysis of Anaesthetic Mishaps*, E. Pierce and J. Cooper, Eds., *International Anaesthesiology Clinics*, Vol. 22, 1984, pp. 167–183.
10. Cooper, J.B., Newbower, R.S., and Kitz, R.J., *An Analysis of Major Errors and Equipment Failures in Anaesthesiology*, Vol. 60, 1984, pp. 34–42.
11. Arcarese, J.S., FDA's Role in Medical Device User Education, *The Medical Device Industry*, N.F., Estrin, Ed., Marcel Dekker Inc., New York, 1990, pp. 129–138.
12. Dripps, R.D., Lamont, A., and Eckenhoff, J.E., The Role of Anaesthesia in Surgical Mortality, *JAMA*, Vol. 178, 1961, pp. 261–266.
13. Clifton, B.S. and Hotten, W.I.T., Deaths Associated with Anaesthesia, *Br. J. Anaes.*, Vol. 35, 1963, pp. 250–259.
14. Edwards, G., Morton, H.J.V., and Pask, E.A., Deaths Associated with Anaesthesia: Report on 1000 cases, *Anaesthesia*, Vol. 11, 1956, pp. 194–200.
15. Cooper, J.B., Newbower, R.S., Charlene, D.L., Long, M.S., and Bucknam, M., Preventable Anaesthesia Mishaps, *Anaesthesiology*, Vol. 49, 1978, pp. 399–406.
16. Weinger, M.B. and Englund, C.E., Ergonomic and Human Factors Affecting Anaesthetic Vigilance and Monitoring Performance in the Operating Room Environment, *Anaesthesiology*, Vol. 75, 1990, pp. 995–1021.
17. Chopra, V., Bovill, J.G., Spierdijk, J., and Koornneef, F., Reported Significant Observations During Anaesthesia: A Prospective Analysis Over an 18-Month Period, *Br. J. Anaes.*, Vol. 68, 1992, pp. 13–17.
18. Cook, R.I. and Woods, D.D., Operating at the Sharp End: The Complexity of Human Error, *Human Error in Medicine*, M.S. Bogner, Ed., Lawrence Erlbaum Associates Publishers, Hillsdale, NJ, 1994, pp. 225–309.
19. Craig, J. and Wilson, M.E., A Survey of Anaesthetic Misadventures, *Anaesthesia*, Vol. 36, 1981, pp. 933–936.
20. Keenan, R.L. and Boyan, P., Cardiac Arrest Due to Anaesthesia, *JAMA*, Vol. 253, 1985, pp. 2373–2377.
21. Couch, N.P., Tilney, N.L., Rayner, A.A., and Moore, F.D., The High Cost of Low-Frequency Events, *N. E. J. Med.*, Vol. 304, 1981, pp. 634–637.
22. Bogner, M.S., Medical Devices: A New Frontier for Human Factors, *CSERIAC Gateway*, Vol. IV, No. 1, 1993, pp. 12–14.
23. Leape, L.L., The Preventability of Medical Injury, *Human Error in Medicine*, M.S. Bogner, Ed., Lawrence Erlbaum Associates, Publishers, Hillsdale, NJ, 1994, pp. 13–25.

24. Keller, J.P., Human Factors Issues with Surgical Devices, *Proceedings of the First Symposium on Human Factors in Medical Devices*, 1989, pp. 34–36.
25. Olivier, D.P., Engineering Process Improvement Through Error Analysis, *Med. Dev. Diag. Ind. Mag.*, Vol. 21, No. 3, 1999, pp. 194–202.
26. Belkin, L., Human and Mechanical Failures Plague Medical Care, *The New York Times*, March 31, 1992, pp. B1 and B6.
27. Casey, S., *Set Phasers on Stun: And Other True Tales of Design Technology and Human Error*, Aegean Inc., Santa Barbara, CA, 1993.
28. Sawyer, D., Do It by Design: Introduction to Human Factors in Medical Devices, Center for Devices and Radiological Health (CDRH), Food and Drug Administration (FDA), Washington, DC, 1996.
29. Wikland, M.E., *Medical Device and Equipment Design*, Interpharm Press Inc., Buffalo Grove, IL, 1995.
30. Bindler, R. and Boyne, T., Medication Calculation Ability of Registered Nurses, *Image*, Vol. 23, 1991, pp. 221–224.
31. Krueger, G.P., Fatigue, Performance, and Medical Error, *Human Error in Medicine*, M.S. Bogner, Ed., Lawrence Erlbaum Associates Publishers, Hillsdale, NJ, 1994, pp. 311–371.
32. Squires, S., Cholesterol Guessing Games, *The Washington Post*, March 6, 1990.
33. Brueley, M.E., Ergonomics and Error: Who is Responsible?, *Proceedings of the First Symposium on Human Factors in Medical Devices*, 1989, pp. 6–10.
34. Hyman, W.A., Human Factors in Medical Devices, *Encyclopaedia of Medical Devices and Instrumentation*, Webster, J.G., Ed., Vol. 3, John Wiley & Sons, New York, 1988, pp. 1542–1553.
35. Le Cocq, A.D., Application of Human Factors Engineering in Medical Product Design, *J. Clin. Eng.*, Vol. 12, No. 4, 1987, pp. 271–277.
36. Countinho, J.S., Failure-Effect Analysis, Transactions of the New York Academy of Sciences, Vol. 26, 1964, pp. 564–84.
37. MIL-STD-1629, Procedures for Performing a Failure Mode, Effects, and Criticality Analysis, Department of Defense, Washington, DC, 1979.
38. MIL-STD-1629A/Notice 2, Procedures for Performing a Failure Mode, Effects, and Criticality Analysis, Department of Defense, Washington, DC, 1984.
39. Dhillon, B.S., Failure Mode and Effects Analysis: Bibliography, *Microelectronics and Reliability*, Vol. 32, 1992, pp. 719–731.
40. Meister, D., Comparative Analysis of Human Reliability Models, Report No. AD 734432, 1971. Available from the National Technical Information Service, Springfield, VA.
41. Dhillon, B.S., *Human Reliability: With Human Factors*, Pergamon Press Inc., New York, 1986.
42. Dhillon, B.S. and Singh, C., *Engineering Reliability: New Techniques and Applications*, John Wiley & Sons Inc., New York, 1981.
43. Wiklund, M.E., Making Medical Device Interfaces More User-Friendly, *Med. Dev. Diag. Ind. Mag.*, Vol. 20, No. 5, 1998, pp. 177–186.

5 Medical Device Software Quality Assurance and Reliability

CONTENTS

5.1 INTRODUCTION

In computer-controlled products, both hardware and software must work normally for their overall success. More specifically, in such products the reliability of software is as important as that of the hardware. Ever since the early days of the computer, the cost of developing software has increased alarmingly compared with developing hardware. For example, in 1955 the software cost accounted for approximately 18% of the total development cost of a software product; in 1985 it increased to 82%.[1,2] This reversal has further increased the importance of software reliability. In fact, with respect to cost, it is estimated that detecting and correcting a software error in normal use is approximately 100 times costlier than detecting and correcting it during the design phase.[1,2]

Software is an important component of medical devices. In 1984, it was estimated that about 80% of major medical systems contained one or more computerized components, and it was predicted that nearly all electrical medical devices would be comprised of at least one computerized part by 1990.[3-5] Furthermore, the size of software in medical devices has increased dramatically. For example, a cardiac rhythm management device consists of approximately 500,000 lines of code.[6] As the result of more computerization of medical devices, the problem of software has become more crucial. In fact, after reviewing the reasons for the recalls of medical devices for the period of 1983–1985 the Food and Drug Administration (FDA) stated that 41 product recalls were related to software; out of which 37 were due to software errors designed into the program prior to the production phase.[2] Furthermore, it may be added that software failures in medical devices have caused deaths and the potential for more such hazards is growing with the complexity of software.[7]

This chapter presents various different aspects of medical device software quality assurance and reliability.

5.2 SOFTWARE TERMS AND DEFINITIONS

There are many terms and definitions associated with software; some of them are as follows:[8,9]

- **Software**. This is computer programs, procedures, rules, and relevant documentation in contrast to physical equipment.
- **Software error**. This is a conceptual, syntactic, or clerical discrepancy that results in one or more faults in the software.
- **Debugging**. This is the process of correcting and isolating errors.
- **Software testing**. This is to determine (i.e., through testing) if a program satisfies specified requirements.
- **Software quality**. This is the totality of characteristics and features of a software item that determines its capability to meet specified requirements or conform to specifications.
- **Software reliability**. This is the probability a given software will operate for a specified time interval, without an error, when used within the framework of designed conditions on the stated machine.
- **Medical device (software)**. Under the U.S. Food, Drug, and Cosmetic Act, any software that satisfies the definition of a medical device is considered a device to which all relevant FDA medical device statutory and regulatory provisions are applicable.[10,11]

5.3 HARDWARE VS. SOFTWARE AND EXAMPLES OF SOFTWARE FAILURES IN MEDICAL DEVICES

Software differs from hardware in many ways. Some of the comparisons between hardware and software are given in Table 5.1.[9,12] Nonetheless, note that medical

TABLE 5.1
Some Comparisons Between Hardware and Software

No.	Hardware	Software
1.	Parts degrade.	Parts do not degrade.
2.	Processes are governed by the laws of physics.	Processes are not governed by the laws of physics.
3.	Failure modes differ from those of software.	Failure modes differ from those of hardware.
4.	Interfaces are physical.	Interfaces are conceptual.
5.	Relatively fewer distinct paths to examine where failures could be hidden.	Relatively more distinct paths to examine where failures could be hidden.
6.	Relatively less sensitive to small errors.	Very sensitive to small errors.

devices possess different software, hardware, and firmware components that are strongly coupled.[13]

Over the years, there have been many medical device failures due to software errors. Some of those failures include[4,5,14-16]

- Two patients died and a third was severely injured due to software errors in a computer-controlled therapeutic radiation machine called Therac 25.[5,17,18] The patients received approximately 125 times the normal dose of 200 rads.[17] A subsequent investigation identified many causes of these incidents, including unjustifiable confidence in software, poor software engineering practices and procedures, and removal of independent hardware safety features. Thus, a single failure in the software caused an accident, and unrealistic and unfinished risk assessment.[19,20]
- Two people died in incidents involving heart pacemakers because of software errors.[5]
- A newly installed computer-aided dispatch system for the London Ambulance Service failed because of software faults, resulting in the wrong ambulance being sent to an incident.[19]
- An infusion pump delivered at the maximum rate instead of intended rate subsequent to entering certain valid data because of a software error.[5]
- A pacemaker lost stored data when any of the parameters were altered.[4,5,21]
- An artificial intelligence medical system provided the wrong advice, leading to a drug overdose.[5]
- A diagnostic laboratory instrument containing a software error resulted in incorrect reports of patient data.[4,5]
- A medical device wrongly processed negative numbers, causing a failure with specific data.[5]
- A pacemaker locked up with loss of system control. The instant printer buffer was full.[5,21]
- A multiple-patient monitoring system failed to store collected data with the correct patient due to a software fault.[4,5,14,15]

5.4 MEDICAL SOFTWARE CLASSIFICATIONS AND SPECIFICATIONS

The Medical Device Amendments of 1976 to the Federal Food, Drug and Cosmetic Act gave the FDA the authority to assure that medical devices in the U.S. are safe and reliable. Consequently, medical devices were grouped into three categories with respect to regulatory requirements. Similarly, the FDA draft policy of 1987 classified medical device software into three groups:[22,23]

- **Group I**. The software belonging to this group is subject to the Act's general controls concerning areas such as record keeping, manufacturer registration, and misbranding. A typical example of this class of software is a program that computes the composition of infant formula.
- **Group II**. The software that falls under this group is the one for which general controls are unsatisfactory to provide a reasonable degree of assurance with respect to safety and effectiveness, as well as for which performance standards can provide an adequate amount of assurance. An example of such software is a program developed to generate radiation therapy treatment plans.
- **Group III**. The software in this category is the one for which inadequate information is available to assure that general controls and performance standards will provide satisfactory assurance of effectiveness and safety. An example of this class of software could be a program that measures glucose levels and computes and dispenses insulin on the basis of calculations obtained without physician intervention.

Furthermore, under the current policy, the FDA classifies software into two categories: (i) stand-alone software, and (ii) software that is a component, part, or accessory to a device. The stand-alone software could be treated as a medical device, thus subject to all applicable FDA medical device statutory and regulatory provisions.[11] The Federal Food, Drug, and Cosmetic Act defines a medical device as "any machine, apparatus, instrument, implant, implement, contrivance, *in vitro* reagent, or other similar or related article, including any component, part, or accessory that is… intended for application in the diagnosis of disease or other conditions, or in the cure, treatment, mitigation, or prevention of disease… or intended to affect the structure or any function of the human body."[10,11]

Some examples of stand-alone medical software are

- Osteoporosis diagnostic software
- Hospital information systems
- Blood bank software that controls donor deferrals and the release of blood products
- Software that performs analysis of potential therapeutic interventions for a certain patient

Software that is a component, part, or accessory to a device is regulated according to the requirements of the parent device unless it is separately classified. The software as a component/part may simply be described as any software intended to be included as an element of the finished, packaged, and labeled device. Examples of such software can be software used in items such as pacemakers, diagnostic X-ray systems, ventilators, infusion pumps, and magnetic resonance imaging devices.

Similarly, software as an accessory may be described as a unit intended to be attached or used in conjunction with another finished device, although it could be sold or promoted as an independent unit. Some examples of such software are[11]

- Software for computing the rate response for a cardiac pacemaker
- Software for computing the bone fracture risk from bone densitometry data
- Software for converting pacemaker telemetry data
- Software for performing statistical analysis of pulse oximetry data

The first step in the development of a medical device software is the availability of specification. Basically, the specification provides information about the purpose of the software and how it will fulfill it. The design and coding time can be reduced by the detailed specification. Nonetheless, the medical device software specification should include items such as program flow diagrams, functional requirements, module definition and interaction, hardware/software interface, hardware review, self diagnostics and error recovery, initialization, memory map, interrupt handling, and data dictionary.[12]

5.5 FRAMEWORK TO DEFINE SOFTWARE QUALITY ASSURANCE PROGRAMS IN A MEDICAL DEVICE COMPANY

In medical device companies, software use is growing rapidly, and many such companies implement software quality assurance programs to ensure that software products and software development processes conform to specified requirements. Unfortunately, in many cases, these programs are not defined effectively leading to nonuniform support among projects.

This section presents a framework that is successfully used in one medical device company to define its software quality assurance program.[24] Such framework could be quite useful to other medical device companies to establish their software quality assurance programs. The framework is composed of eight steps as shown in Figure 5.1.

The first step involves gathering all information related to the program, including internal and external requirements, organizational culture, and industry standards. The second step establishes a plan that includes factors such as important activities to be accomplished, required resources, and a schedule of expected completion dates. The third step establishes a mission statement of software quality assurance that is traceable to the company's mission statement and approved by the upper-level management.

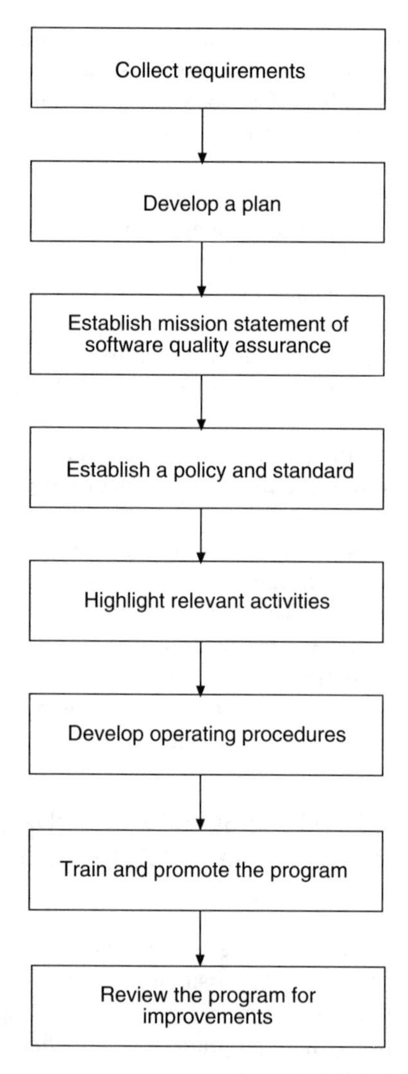

FIGURE 5.1 Framework steps for defining the roles and responsibilities of software quality assurance.

The fourth step develops a policy and a standard. The policy focuses on the important organizational concerns, such as software quality assurance having the authority to perform independent evaluations of software development processes and products, and the authority to take appropriate corrective measures to improve the quality of software products and development processes. The standard provides description supporting the policy. The fifth step identifies specific activities. This includes analysis of the existing activities (if any) being performed, and the identification of the most valuable and additional activities necessary to satisfy the mission statement.

The sixth step establishes standard operating procedures. These procedures provide adequate information to conduct software quality assurance activities. The seventh step conducts appropriate training and promoting the program. The final step of the framework involves evaluating the program for improvements.

5.6 SOFTWARE RELIABILITY ASSURANCE PROGRAMS FOR THE CARDIOPLEGIA DELIVERY SYSTEM, AND IMPORTANT SOFTWARE ISSUES IN THE SAFETY OF CARDIAC-RHYTHM-MANAGEMENT PRODUCTS

Software reliability is important in the cardioplegia delivery system which provides multiple combinations of blood, crystalloid, arresting agent, and additive solutions under the perfusionist's programmable control. During the development of the operating software for the cardioplegia delivery system, the software reliability assurance program's core elements that can be implemented include simplified design, documented performance requirements, code inspections, safety hazard analysis, periodic audits, design and coding standards, extensive testing starting at the module level, quality practices and standards, requirement tracing, clearly defined user interfaces(s), documentation and resolution of all defects, and reliability modeling to predict operational reliability.[25]

With respect to cardiac-rhythm-management products, there are various important issues to consider when analyzing their software safety: marketing issues, technical issues, management issues, etc. Thus, software developers must evaluate such issues in light of product context, constraints, and environmental requirements.[26] Examples of the cardiac-rhythm-management systems are pacemakers and defibrillators used to provide electrical therapy to malfunctioning cardiac muscles.

Marketing issues have four major components: product requirements, market requirements, regulatory requirements, and legal requirements. With respect to the product requirements, a typical cardiac-rhythm-management device is made up of advanced software, electrical subsystems, and mechanical subsystems which must perform effectively for the overall success of the device. In particular, because the software subsystem of modern devices is composed of approximately 500,000 lines of code, its safety, reliability, and efficiency to control both internal and external system operations are of utmost importance.

In regard to the market requirements, the sheer size of the market of cardiac-rhythm-management systems is the sole important factor in their safety. For example, it was predicted that the world market for such devices for 1997 would amount to about $3 billion. Furthermore, each year nearly half a million people experience a sudden cardiac-death episode and approximately 31,000 people receive a defibrillator implant in the U.S. alone.[27] Also, about the same number of implants are performed in the rest of the world each year.[26]

The regulatory requirements are important, as the companies producing cardiac-rhythm-management systems are regulated by agencies such as the FDA. Such regulatory agencies demand systematic and rigorous software-development processes, including safety analysis. It is estimated that roughly half of the development cycle is consumed by the regulatory acceptance process, thus the development procedures used must satisfy regulatory requirements effectively, otherwise there could be long delays.

The legal requirements are another important factor because of the life-or-death nature of these systems/devices, and the concerned regulations lead to highly sensitive legal requirements involving patients and their families, manufacturers, regulatory agencies, etc. More specifically, in the event of death or injury to a patient, lawsuits by the regulatory agencies and the termination of products by these agencies because of their safety problems are the driving factors for the legal requirements.

The technical issues are also a major factor during software development because the software complexity and maintenance concerns determine the analysis method and the incorporation process to be used. With respect to complexity, modern devices can have tens of parallel, real-time, asynchronous software tasks reacting to random external events. Thus, ensuring the correct and timely behavior of such complex software is rather difficult, as is the identification and mitigation of safety faults with correctness and timeliness. In regard to maintenance, the code resulting from the modification of safety-critical software is one of the most critical components. A survey conducted by one developer of medical device software indicates that out of the software-related engineering change requests, and after the internal release of software during development, 41% were related to software safety.[26]

Finally, the management issues are basically concerned with making changes to the development process to incorporate explicit safety analysis, thus requiring a clear vision and convincing justifications. Consequently, management has to be convinced that extra safety-related tasks will help the set goals in the long run despite the perceived additional short-term efforts. It will certainly help if the management can be shown explicitly that the cost of conducting software-safety analysis during the development cycle can help reduce the overall development cost, regulatory problems, and market losses, and avoid future legal costs.

5.7 SOFTWARE DESIGN, CODING, TESTING, AND METRICS

A software life cycle could be divided into many phases, such as, requirement definition, preliminary design, detail design, coding, integration and testing, and maintenance.[9,28,29]

This section discusses the important aspects of software design, coding, testing, and metrics because they are critical to producing safe and reliable medical device software.

5.7.1 Software Design

Software design and implementation may simply be described as a multistaged process that translates system and software requirement into a functional program that addresses each requirement. The basis of good software design is a combination of creativity and discipline, but it is important to note that approximately 45% of software system errors tend to be design errors.[9,30]

Software design may be divided into two distinct categories: top-level design and detailed design.[31] The important elements of the top-level design are methodology selection, language selection, design alternatives and trade-offs, software architecture, object-oriented design, software review, structural analysis, risk analysis, and software requirements traceability matrix. Similarly, the major components of the detailed design include design techniques, module specification, design support tools, performance predictability and design simulation, coding and design as a basis for verification and validation testing.

Design Representation Methods

In software design, many types of design representation schemes are employed. Unfortunately, there is no single commonly used scheme. Nonetheless, some of these schemes are flowcharts, Chapin charts, hierarchy plus input-process-output [HIPO] diagram, Warnier-Orr diagram, data structure chart, and code listing. These schemes are described in detail in references 9 and 32–34.

Software Design Techniques

Various methods are used to help produce better software. Three commonly used methods are shown in Figure 5.2. Each method is described below.

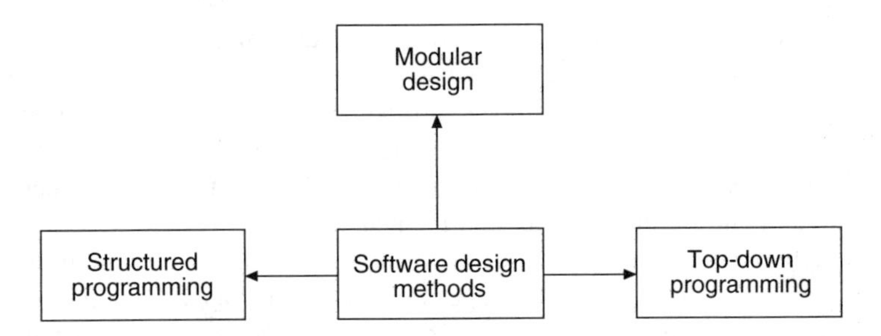

FIGURE 5.2 Commonly used software design techniques.

Modular Design

A module may simply be described as a self-contained, modest-sized subprogram that performs independently on one specific function, and if it is taken away from the system, it will only disable its unique function and nothing else. Thus, a complex medical software development job can be decomposed into various modules. Two major criteria associated with modular design are module size restriction and functional independence restriction.

The module size restriction is concerned with defining the maximum and minimum number of program language statements in a given module. Although such statements may vary from one project to another, reference 35 suggests 50 maximum statements and five minimum statements in a module. The functional independence restriction is concerned with defining the logic scope a module may cover; however, in this case the good judgement of programmers is an important factor.

There are various advantages and disadvantages of the modular design.[9] Advantages include being able to easily write, debug, and test software; having more reliable software; and being able to make changes and error corrections during the deployment phase thereby reducing cost. Major disadvantages are that it may require more memory space, and it definitely requires more effort during design.

Top-Down Programming

Top-down programming is also known as hierarchic decomposition because of the cascading nature of the process.[36] It may be described as a decomposition process that directs attention to the program control structure or to the program flow of control. The method begins with a module representing the total program; then the module is decomposed into subroutines. In turn, the subroutines are broken down further, until the broken down elements are easily comprehensible and straightforward to work with. The analogy of this method may be drawn to a tree, the trunk representing the module and the leaves representing the statements.

Top-down programming has many advantages including increasing confidence in software; presenting the end product in a more readable form; having better quality software; and lowering the cost of testing.

Structured Programming

Since the late 1960s, structured programming has been receiving a considerable amount of attention, and may be described as coding that avoids the use of GO TO statements and program design that is top-down and modular.[37] For an average programmer, the easy-to-use rules associated with structured programming include restricting the use of GO TO statements in modules, ensuring that each module has only one entry and one exit, using a language and compiler that have structured programming capabilities for implementation, and ensuring that all statements with the inclusive of subroutine calls are commented.[35]

Just like any other software design technique, structured programming has advantages and disadvantages. Advantages include the fact that it is useful to localize an error; one can maximize the amount of code that can be reused in redesign effort; it is useful in understanding the program design by the designers and others; and it is an effective tool to increase programmer productivity. The main disadvantages are that it may require more memory and running time, and it requires additional effort because many programming languages lack certain control concepts appropriate to implement the structured programming purpose.

5.7.2 Software Coding

The main goal of the software development implementation phase is to translate design specifications into source code. Thus, the source code and internal documentation are written so that it is simple and straightforward to verify conformance of the code to its specification, in turn, making it easy to debug, test, and modify software. Factors such as structured coding techniques, good coding style, appropriate supporting documents, and good internal comments help enhance clarity of the source code.

In particular, good coding style or practices can be achieved by following guidelines such as reviewing each line of code, requiring coding sign-offs, rewarding good code, emphasizing that code listings are public assets, and routing good code examples for review.[31,38,39] All in all, the implementation errors introduced by software professionals during the implementation phase can lead to infrequent, bizarre, and potentially dangerous outcomes.[40]

5.7.3 Software Testing

Software testing may simply be defined as the process of executing a program with the intention of discovering bugs or errors. The prime factors behind the testing of medical device software are as follows:[41,42]

- Increased concern for safety, as the consequences of poorly tested software could be very devastating to both patients and producers
- Competition
- Regulatory pressure from agencies such as the FDA

Note that testing accounts for a substantial proportion of the software development process. Thus, to have an effective testing program, it is advisable to develop a good test plan by considering factors such as duration of the text phase, functions needed to be tested, test result documentation, test strategy, error classification scheme, performance requirements, and the testing criteria to be satisfied.[43]

Software testing may be classified into two groups: manual testing and automated testing.[41] Both groups are described below.

Manual Software Testing

There are many traditional approaches to manual software testing; some are as follows:[42]

- Error testing
- Safety testing
- Functional testing
- White-box testing
- Free-form testing

Error testing involves ensuring that the device performs correctly in abnormal situations such as internal parts failures, power surges or outages, or incomprehensible values generated by peripheral devices. Safety testing concerns focusing test efforts on conditions potentially harmful to the patient or user. Functional testing is an orderly process that directs test efforts on what the device is expected to do. Because this kind of testing forms the heart of traditional validation and verification procedures, it normally receives the most attention and the greatest number of test cases, since each function is tested exhaustively. White-box testing is employed when it is required to verify the internal functioning of the device software. This type of testing allows the tester to look inside the device and develop tests to discover weak spots in internal logic of the program. Furthermore, it is possible to incorporate white-box methods into error testing, safety testing, functional testing, and free-form testing.

Free-form testing is used because no formal test plan can find each important software bug. Thus, this type of testing subjects the device to additional stress on top of the formal approaches described above. Free-form testing may also be described as a process in which testers with a varying degree of knowledge use the device in unconventional and unexpected ways in an attempt to provoke failures. Although this type of testing normally finds a large percentage of bugs, there is no assurance that it will cover all potential problem areas because of its unstructuredness.[42]

Automated Software Testing

Automated software testing uses computer technology to test software so that the test process is faster, more accurate, more complete, and is backed up with better documentation. More specifically, automated software test systems employ computer technology for purposes such as to stimulate the test target, monitor its response, record results, and control the complete process.

In the case of stimulating the test target, automated testing augments environmental, human, and internal stimulation of the target through computer simulations. With respect to monitoring its response, automated testing augments human test monitoring through computerized monitoring systems. With record results, automated testing replaces handwritten test-related notes with computer-based records. In the case of controlling the complete process, automated testing completely replaces manual test plans with computer-based test programs.

5.7.4 Software Metrics

Since the mid 1970s, software metrics have been developed and used to measure software complexity. A metric may be described as quantified figure-of-merit for useful accuracy and reliability. The term is customarily applied to software.[8] To obtain a true indication of quality and reliability of a given software, it must be subjected to measurement and the quality attributes. They, in turn, must be associated to specific and quantifiable device or product requirements. This can be accomplished by using metrics. There are many uses of metrics, including goal setting, project planning and managing, improving quality and productivity, and enhancing customer confidence.[12,31]

During the selection of metrics, a careful consideration must be given to factors such as usefulness to the specific objectives of the program, derivation from the program requirements, and the determination of the software in line with the given requirements. Because the metric computation is subject to the availability of data, such data must be fairly accurate, consistent, and collected from many projects. Two commonly used software metrics are described below.

McCabe's Complexity

The basis for this metric is the complexity of the flow-of-control in the system. More specifically, the metric relates the complexity to the software or program measure of structure. The metric is derived from classical graph theory and is expressed by the following equation:[12,31,44-46]

$$N_C = \alpha - \theta + 2x \tag{5.1}$$

where

N_C	is the complexity number.
x	is the number of connected components or separate tasks.
α	is the total number of edges in the program under consideration.
θ	is the number of vertices or nodes.

Example 5.1

Calculate by assuming $x = 2$, $\alpha = 4$, and $\theta = 7$ in Equation (5.1), the value of the complexity number. Comment on the result.

By substituting the given data into Equation (5.1) we get

$$N_C = 4 - 7 + 2(2)$$

$$N_C = 1$$

As the value of the complexity number is equal to one, it means that high software reliability can be expected. All in all, it may be said that the higher the value of the complexity number, the more difficult it will be to build, test and maintain software, thus lowering its reliability.

Halstead Measures

These metrics are due to Professor Maurice Halstead of Purdue University[47-49] and provide a quality measure of the software development process by considering two factors: (i) number of distinct operands (i.e., variables and constraints) and (ii) number of distinct operators (i.e., types of instructions).[12,31] Professor Halstead defined various characteristics that can be quantified about the given software and then associated them in various ways to explain the different aspects of the software under consideration. Halstead measures include program length, software vocabulary, software volume, and the potential volume.[47-49] Each measure is described below.

Program Length

This is expressed by

$$L = x + y, \tag{5.2}$$

where

L is the length of the program.
x is the total occurrences of operators.
y is the total occurrences of operands.

Software Vocabulary

This is defined by

$$M = X + Y, \tag{5.3}$$

where

M is the vocabulary of the software.
X is the number of distinct operators.
Y is the number of distinct operands.

Software Volume

This is given by

$$V = L \log_2 M, \tag{5.4}$$

where

V is the volume of the software

Potential Volume

This is defined by

$$V^* = (2 + Y) [\log_2 (2 + Y)], \tag{5.5}$$

where

V^* is the potential volume.

5.8 SOFTWARE RELIABILITY MODELING FOR MEDICAL DEVICES

The history of software reliability may be traced back to 1964 with the construction of a histogram of problems per month concerning switching system software.[50] In 1967, G.R. Hudson and R.W. Floyd proposed Markov birth-death models and approaches for formal validation of software programs, respectively.[51,52] Since then, many software reliability models have been developed.[53-55] A list of references as of 1987 on software reliability is given in reference 9.

This section presents two software reliability models considered useful for medical devices.

USAF Model

This model was developed by the U.S. Air Force Laboratory in Rome, NY[56] to predict software reliability during the initial phases of the software life cycle. The model begins by developing the predictions of fault density and then transforming them into other reliability measures such as failure rates.[57] Thus, this initial fault density is expressed as follows:[56,57]

$$IFD = \sum_{i=1}^{11} \theta_i , \tag{5.6}$$

where

IFD is the initial fault density.

θ_i is the ith factor related to fault density at the initial phases of software; i = 1 (program size), i = 2 (complexity), i = 3 (nature of application), i = 4 (traceability), i = 5 (development environment), i = 6 (language type), i = 7 (modularity), i = 8 (extent of reuse), i = 9 (quality review results), i = 10 (standards review results), and i = 11 (anomaly management).

The initial failure rate is defined by

$$\lambda_i = kmR\,(IFD), \tag{5.7}$$

where

λ_i is the initial failure rate.

R is the fault expose ratio ($1.4 \times 10^{-7} \leq R \leq 10.6 \times 10^{-7}$)

k is the number of lines of source code.

m is the program's linear execution frequency.

The linear execution frequency of the program is given by

$$m = \beta/\alpha, \tag{5.8}$$

where

β is the average instruction rate.

α is the number of object instructions in the program.

In turn, the object instructions in the program are expressed by

$$\alpha = (SI)r, \tag{5.9}$$

where

r is the code expansion ratio (i.e, the ratio of machine instructions to source instructions and usually, its mean value is taken as 4.)

SI is the number of source instructions.

The inherent faults are defined by

$$F_i = k\,(IFD), \tag{5.10}$$

where

F_i is the number of inherent faults.

Thus, by substituting Equations (5.8–5.10) into Equation (5.7), we get

$$\lambda_i = \beta R F_i/(SI)r. \tag{5.11}$$

Mills Model

This model was developed by H.D. Mills[58] by reasoning that an estimation of faults remaining in a given software could be made through a seeding process that assumes a homogeneous distribution of a representative class of faults. Before starting the seeding process, fault analysis is performed to determine the expected types of faults in the code and their relative frequency of occurrence.

During review or testing, both seeded and unseeded faults are identified. In turn, these highlighted faults are used to determine the number of remaining faults for the fault type under consideration. Nonetheless, the calculation of this measure is subject to the discovery of the faults. More specifically, in the event of zero discovered faults, this measure cannot be computed.

The maximum likelihood of the unseeded faults is defined by[59]

$$F_u = F_s f_u / f_s, \tag{5.12}$$

where

F_u is the maximum likelihood of the unseeded faults.
F_s is the number of seeded faults.
f_s is the number of seeded faults discovered.
f_u is the number of unseeded faults uncovered.

The number of unseeded faults remaining in the software is given by

$$F = F_u - f_u. \tag{5.13}$$

Example 5.2

A software was seeded with 40 faults, and during the testing process 65 faults of the same kind were uncovered, i.e., 25 seeded faults and 40 unseeded faults. Calculate the number of unseeded faults still remaining in the software under consideration.

By substituting the given data into Equation (5.12), we get

$$F_u = (40)(40)/25 = 64 \text{ faults}$$

Using the above result and the remaining given data Equation (5.13) yields

$$F = 64 - 40 = 24 \text{ faults}$$

It means there are still 24 unseeded faults in the software.

5.9 PROBLEMS

1. Define the following terms:
   - Software reliability
   - Software error
   - Software quality
2. Discuss at least five medical device failures due to software.
3. Make a comparison between hardware and software.
4. Describe three FDA medical device software classifications.
5. Give at least four examples of the "stand-alone medical software" as defined by the FDA.
6. What are the important software issues in the safety of cardiac-rhythm-management products?
7. Describe the following two software design techniques:
   - Structured programming
   - Modular design
8. Describe the followings two items:
   - Manual software testing
   - Automated software testing
9. Discuss the following two terms:
   - Software metrics
   - Software coding
10. A computer program was seeded with 50 faults, and after intensive testing a total of 75 faults were discovered. A careful examination of these faults revealed 50 seeded faults and 25 unseeded faults. Estimate the number of unseeded faults remaining in the program.

REFERENCES

1. Boehm, B.W., Software Engineering, *IEEE Transactions on Computers*, Vol. 25, 1976, pp. 1226–1235.
2. Estrin, N.F., Ed., *The Medical Device Industry*, Marcel Dekker Inc., New York, 1990.
3. Anbar, M., Ed., *Computers in Medicine*, Computer Science Press, Rockville, MD, 1987.
4. Bassen, H., Silberberg, J., Houston, F., Knight, W., Christman, C., and Greberman, M., Computerized Medical Devices: Usage trends, problems, and safety technology, *Proceedings of the 7th Annual Conference of the IEEE/Engineering in Medicine and Biology Society on Frontiers of Engineering and Computing in Health Care*, 1985, pp. 180–185.

5. Schneider, P. and Hines, M.L.A., Classification of Medical Software, *Proceedings of the IEEE Symposium on Applied Computing*, 1990, pp. 20–27.
6. Vishnuvajjala, R.V., Subramaniam, S., Tsai, W.T., Elliott, L., and Mojdehbaksh, R., Run-Time Assertion Scheme for Safety-Critical Systems, *Proceedings of the 9th IEEE Symposium on Computer-Based Medical Systems*, 1996, pp. 18–23.
7. Wood, B.J. and Ermes, J.W. Applying Hazard Analysis to Medical Devices, Part II: Detailed Hazard Analysis, *Med. Dev. Diag. Indust. Mag.*, Vol. 15, No. 3, 1993, pp. 58–64.
8. Omdahl, T.P., *Reliability, Availability, and Maintainability (RAM) Dictionary*, ASQC Quality Press, Milwaukee, 1988.
9. Dhillon, B.S., *Reliability in Computer System Design*, Ablex Publishing Corporation, Norwood, NJ, 1987.
10. Federal Food, Drug, and Cosmetic Act, as Amended, Sec. 201 (h), U. S. Government Printing Office, Washington, DC, 1993.
11. Onel, S., Draft Revision of FDA's Medical Device Software Policy Raises Warning Flags, *Med. Dev. Diag. Indust. Mag.*, Vol. 19, No. 10, 1997, pp. 82–91.
12. Fries, R.C., *Reliability Assurance for Medical Devices, Equipment and Software*, Interpharm Press Inc., Buffalo Grove, IL, 1991.
13. Subramanian, S., Elliott, L., Vishnuvajjala, R.V., Tsai, W.T., and Mojdehbakhsh, R., Fault Mitigation in Safety-Critical Software Systems, *Proceedings of the 9th IEEE Symposium on Computer-Based Medical Systems*, 1996, pp. 12–17.
14. Neumann, P.G., Risks to the Public, ACM SIGSOFT Software Engineering Notes, Vol. 11, No. 2, 1986.
15. Neumann, P.G., Some Computer-Related Disasters and Other Egregious Horrors, *Proceedings of the 7th Annual Conference of the IEEE/Engineering in Medicine and Biology Society*, 1985, pp. 1238–1239.
16. Smith, C.E. and Peel, D., Safety Aspects of the Use of Micro-Processors in Medical Equipment, *Measurement and Control (U.K.)*, Vol. 21, No. 9, 1988, pp. 275–276.
17. Gowen, L.D. and Yap, M.Y., Traditional Software Development's Effects on Safety, *Proceedings of the 6th Annual IEEE Symposium on Computer-Based Medical Systems*, 1993, pp. 58–63.
18. Joyce, E., Software Bugs: A Matter of Life and Liability, *Datamation*, Vol. 33, No. 10, 1987, pp. 88–92.
19. Shaw, R., Safety-Critical Software and Current Standards Initiatives, *Computer Methods and Programs in Biomedicine*, Vol. 44, 1994, pp. 5–22.
20. Leveson, N.G. and Turner, C.S., An Investigation of the Therac-25 Accidents, *IEEE Comput.*, July 1993.
21. Bonnett, B.J.M., Software System Safety, *Proceedings of the 7th Annual Conference of the IEEE/Engineering in Medicine and Biology Society*, 1985, pp. 186–192.
22. FDA Policy for the Regulation of Computer Products (Draft), Food and Drug Administration (FDA), Washington, DC, 1987.
23. Estrin, N.F., Ed., *The Medical Device Industry*, Marcel Dekker Inc., New York, 1990.
24. Linberg, K.R., Defining the Role of Software Quality Assurance in a Medical Device Company, *Proceedings of the 6th Annual IEEE Symposium on Computer-Based Medical Systems*, 1993, pp. 278–283.
25. Heydrick, L., Jones, K.A., Applying Reliability Engineering During Product Development, *Med. Dev. Diag. Indust. Mag.*, Vol. 18, No. 4, 1996, pp. 80–84.
26. Mojdehbakhsh, R., Tsai, W.T., Kirani, S., and Elliott, L., Retrofitting Software Safety in an Implantable Medical Device, *IEEE Software*, Vol. 11, January 1994, pp. 41–50.

27. Lowen, B., Cardiovascular Collapse and Sudden Cardiac Death, in *Heart Disease Textbook of Cardiovascular Medicine*, W. B. Saunders Co., Philadelphia, 1984, pp. 778–808.
28. Fisher, K.F., A Methodology for Developing Quality Software, *Proceedings of the Annual American Society for Quality Control Conference*, 1979, pp. 364–371.
29. Dunn, R. and Ullman, R., *Quality Assurance for Computer Software*, McGraw-Hill Book Company, New York, 1982.
30. Joshi, R.D. Software Development for Reliable Software Systems, *J. Sys. Soft.*, Vol. 3, 1983, pp. 107–121.
31. Fries, R.C., *Reliable Design of Medical Devices*, Marcel Dekker Inc., New York, 1997.
32. Fairly, R.E., Modern Software Design Technique, *Proceedings of the Symposium on Computer Software Engineering*, 1976, pp. 111–131.
33. Peters, L.J. and Tripp, L.L., Software Design Representation Schemes, *Proceedings of the Symposium on Computer Software Engineering*, 1976, pp. 31–56.
34. Shooman, M.L., *Software Engineering: Design, Reliability, and Management*, McGraw-Hill Book Company, New York, 1983.
35. Wang, R.S., Program with Measurable Structure, *Proceedings of American Society for Quality Control Conference*, 1980, pp. 389–396.
36. Koestler, A., *The Ghost in the Machine*, MacMillan, New York, 1967.
37. Rustin, R., *Debugging Techniques in Large Systems*, Prentice Hall Inc., Englewood Cliffs, NJ, 1971.
38. McConnell, S.C., *Code Complete: A Practical Handbook of Software Construction*, Microsoft Press, Redmond, WA, 1993.
39. Maguire, S.A., *Writing Solid Code: Microsoft's Techniques for Developing But-Free C Programs*, Microsoft Press, Redmond, WA, 1993.
40. Levkoff, B., Increasing Safety in Medical Device Software, *Med. Dev. Diag. Indust. Mag.*, Vol. 18, No. 9, 1996, pp. 92–101.
41. Jorgens, J., The Purpose of Software Quality Assurance: A Means to an End, in *Developing Safe, Effective, and Reliable Medical Software*, Association for the Advancement of Medical Instrumentation, Arlington, VA, 1991, pp. 1–6.
42. Weide, P., Improving Medical Device Safety with Automated Software Testing, *Med. Dev. Diag. Indust. Mag.*, Vol. 16, No. 8, 1994, pp. 66–79.
43. Kopetz, H., *Software Reliability*, the MacMillan Press Ltd., London, 1979.
44. McCabe, T., Notes on Software Engineering, 5380 Mad River Lane, Columbia, MD 21044, 1975.
45. McCabe, T., A Complexity Measure, *IEEE Transactions on Software Engineering*, Vol. 2, No. 4, 1976, pp. 308–320.
46. Shooman, M.L., *Software Engineering*, McGraw-Hill Book Company, New York, 1983.
47. Halstead, M., Software Physics: Basic Principles, Report No. R.J. 1582, IBM Research Laboratory, Yorktown Heights, NY, May 20, 1975.
48. Halstead, M., A Quantitative Connection Between Computer Programs and Technical Prose, Digest of Papers, COMPCON'77, IEEE, New York, September 1977, pp. 332–335.
49. Halstead, M.H., *Elements of Software Science*, Elsevier North-Holland Inc., New York, 1977.
50. Haugk, G., Tsiang, S.H., and Zimmerman, L., System Testing of the No. 1 Electronic Switching System, *Bell System Tech. Journal*, Vol. 43, 1964, pp. 2575–2592.
51. Hudson, G.R., Programming Errors as a Birth-and-Death Process, Report No. SP-3011, System Development Corporation, Santa Monica, CA, 1967.

52. Floyd, R.W., Assigning Meanings to Program, *Mathematical Aspects of Computer Science*, Vol. X1X, 1967, pp. 19–32.
53. Musa, J.D., Iannino, A., and Okumoto, K., *Software Reliability*, McGraw-Hill Book Company, New York, 1987.
54. Sukert, A.N., An Investigation of Software Reliability Models, *Proceedings of the Annual Reliability and Maintainability Symposium*, 1977, pp. 478–484.
55. Schick, G.J. and Wolverton, R.W., An Analysis of Competing Software Reliability Models, *IEEE Transactions on Software Engineering*, Vol. 4, 1978, pp. 104–120.
56. Methodology for Software Reliability Prediction and Assessment, Report No. RL-TR-92-52, Volumes I and II Rome Air Development Center, Griffiss Air Force Base, Rome, NY, 1992.
57. Lyn, M.R., Ed., *Handbook of Software Reliability Engineering*, McGraw-Hill Companies Inc., New York, 1996.
58. Mills, H.D., On the Statistical Validation of Computer Programs, Report No. 72-6015, IBM Federal Systems Division, Gaithersburg, MD, 1972.
59. Pecht, M., Ed., *Product Reliability, Maintainability, and Supportability Handbook*, CRC Press, Boca Raton, FL, 1995.

6 Medical Device Safety Assurance

CONTENTS

6.1 INTRODUCTION

A medical device must not only be reliable but also safe. More specifically, a medical device must never operate or fail in a way that can be harmful to the patient or the user. The problem of safety concerning humans is not new; it can be traced back to the ancient Babylonians. In 2000 B.C., ruler Hammurabi developed his code known as the "Code of Hammurabi" with respect to health and safety. It contained clauses concerning injuries and financial damages against those causing injury to others.[1,2]

The Romans were also very concerned with health and safety as depicted by their construction projects. In particular, they built well-ventilated houses, sewerage systems, aqueducts, etc.

In modern times, the passage of the Occupational Safety and Health Act (OSHA) in 1970 in the U.S. is considered as an important milestone with respect to health and safety. Two other important milestones specifically concerning medical devices, in the U.S., are the Medical Device Amendments of 1976, and the Safe Medical Device Act (SMDA) in 1990.

This chapter presents many important aspects associated with medical device safety assurance.

6.2 MEDICAL DEVICE SAFETY VS. RELIABILITY

Although safety and reliability are good things to which all medical devices should aspire, often there is some confusion, particularly in the industrial sector, concerning the difference between device safety and reliability. Nonetheless, it may be added that safety and reliability are distinct concepts and at times can have conflicting concerns.[3]

A safe medical system may be described as a system that does not cause too much risk to people, equipment, or property. In turn, risk is an undesirable event that can occur and is measured with respect to severity and probability. More specifically, device safety is simply a concern with failures that introduce hazards and is expressed in terms of the level of risk, not in terms of meeting requirements. In contrast, reliability of a medical device is the probability of success to meet its requirements.

A medical device is still considered safe even if it often fails without leading to mishaps. On the other hand, if a device operates normally at all times but regularly puts individuals at risk, such device is considered reliable but unsafe. The following are examples of reliable/unsafe and safe/unreliable medical devices, respectively:[3]

- A pacemaker that can pace, say at 110 beats per minute under any condition, is considered very reliable. However, if the patient involved is in cardiac failure, a high pacing rate is considered, from a medical standpoint, inappropriate. Under this scenario the pacemaker is reliable but unsafe.
- A pacemaker that does not always pace at the programmed rate for a small number of patients is not a safety concern. Under this scenario the pacemaker is safe but unreliable.

6.3 TYPES OF MEDICAL DEVICE SAFETY AND MEDICAL DEVICE HARDWARE AND SOFTWARE SAFETY

Medical device safety may be classified into three categories as shown in Figure 6.1.[4] Unconditional safety is preferred over all other possibilities because it is most

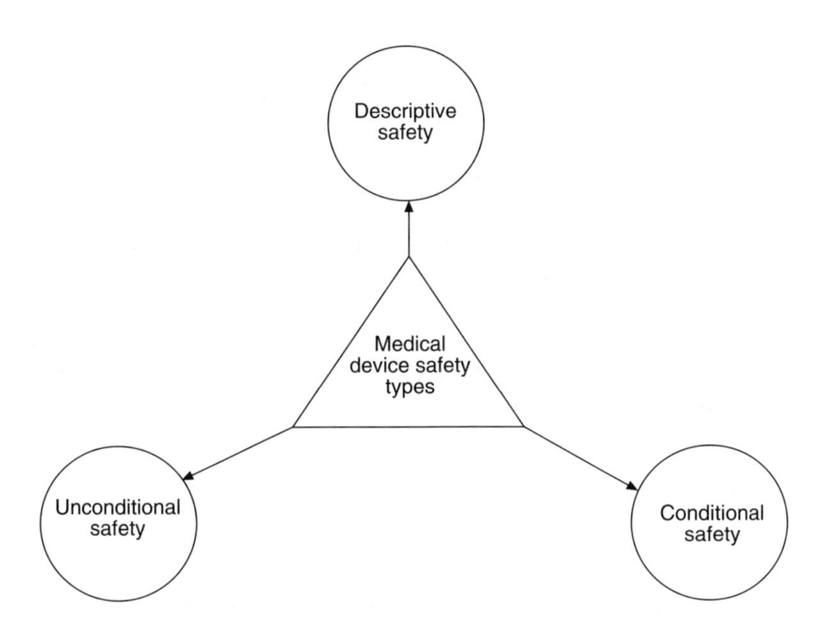

FIGURE 6.1 Types of medical device safety.

effective. However, it requires eradication of risks associated with devices by design. Furthermore, it is to be noted that the use of warnings complements satisfactory design but does not replace it.

Conditional safety is used when it is impossible to realize unconditional safety. For example, in the case of on X-ray or laser surgical device, it is not feasible to avoid emissions of dangerous radiation. However, it is within means to minimize risk with actions such as limiting access to therapy rooms or including a locking switch that allows device activation by authorized individuals only. Additional, indirect safety means are X-ray folding screans, protective laser glasses, etc.

The conditions for safe operation of the device are defined or described when the conditional safety does not lead to the intended results. More specifically, the descriptive safety is used only when it is not possible or appropriate to provide safety by conditional or unconditional means.

Descriptive safety with respect to mounting, transport, operation, connection, maintenance, and replacement may simply be statements such as "This side up," "Handle with care," "Not for explosive zones," etc.

Device hardware safety is important because many parts such as electronic parts are vulnerable to factors such as environmental stresses and electrical interferences. This requires that each part in a medical device be analyzed with respect to potential failures and safety concerns. For this purpose, there are several methods and techniques available to the analyst: failure modes and effect analysis (FMEA), fault tree analysis, and so on. Subsequent to part/component analysis, approaches such as safety margin, load protection, and component derating can be used to lower the potential for the failure of components identified as critical.

Safety of software is equal in importance to that of the hardware elements. However, it may be said that software in and of itself is not unsafe, but the physical systems it may control can cause damage. Consequences of a software failure in medical devices could be quite serious. For example, an out of control program can drive a radiation therapy machine gantry into a patient or a hung program may not only fail to stop a radiation exposure but also deliver an overdose.[3]

The problem of software safety in medical devices is quite serious, as highlighted by a Food and Drug Administration (FDA) study concerning "device recalls." This study conducted over a period of five years (from 1983–1989) indicates that a total of 116 problems in software quality resulted in the recall of medical devices in the U.S. All in all, many methods and techniques can be used to improve software safety in medical devices.[5]

6.4 ESSENTIAL SAFETY REQUIREMENTS FOR MEDICAL DEVICES AND LEGAL ASPECTS OF DEVICE SAFETY

Over the years, the government and other agencies have placed various types of requirements on medical devices directly or indirectly with respect to safety. These requirements may be grouped into three areas: safe design, safe function, and sufficient information.[4] Seven requirements belonging to the safe design area are shown in Figure 6.2.

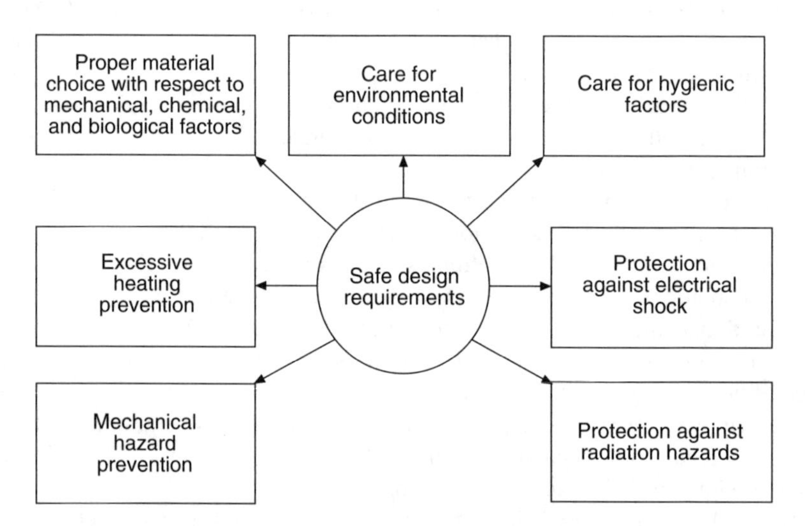

FIGURE 6.2 Requirements related to safe medical device design.

Although the requirements shown in Figure 6.2 are considered self-explanatory, some are described in a little more detail. The care for environmental conditions includes factors such as humidity, electromagnetic interactions, and temperature. Mechanical hazard prevention includes factors such as breaking strength, safe distances, and stability of the device requiring a careful attention. Mechanisms for the prevention of excessive heating are temperature control, cooling, and effective design.

The elements of the safe function group are reliability, warning for or prevention of dangerous outputs, and accuracy of measurements.

The components of the sufficient information group include effective labeling, instructions for use, accompanying documentation, production, and packaging.

One of the objectives of system safety is to limit legal liability. Tort law complements safety regulation through the deterrent of producing harmful medical devices, along with its fundamental objective of providing compensation to injured people. Recently, the U.S. Supreme Court's decision on Medtronic, Inc. vs. Lohr has put additional pressure on manufacturers to produce safe and reliable medical devices. The case was filed by Lora Lohr, a Florida woman, who was implanted with a cardiac pacemaker manufactured by Medtronic, Inc. to correct her abnormal heart rhythm. The pacemaker failed, and she alleged that its failure was due to defective design, manufacturing, and labeling.[6] In some areas of the medical device industry, the product liability problem has already increased the cost of conducting business.

The following three theories are commonly used to make manufacturers liable for injury caused by their products:

- Negligence
- Strict liability
- Breach of warranty

In the case of negligence, if the manufacturer of a device fails to exercise reasonable care or fails to satisfy a reasonable standard of care during the device manufacture, handling, or distribution, it could be liable for any damages caused by the device. The basis for imposing strict liability is that the device manufacturer is in the best position to lower associated risks. More specifically, it means that the device manufacturer, even if not negligent, is still responsible for putting its product on the open market.

The breach of warranty may be alleged under three scenarios: breach of an expressed warranty, breach of the implied warranty of suitability for a specific case, and breach of the implied warranty of merchantability. For example, if a device caused injury to an individual because of its failure to operate as warranted, the manufacturer of that device faces liability under the first scenario (i.e., breach of an expressed warranty).

6.5 SAFETY IN DEVICE LIFE CYCLE*

To have a safe medical device, safety has to be considered throughout its life cycle. Thus, the life cycle of a medical device may be divided into five distinct phases as follows:[7]

* The safety-related activities outlined in this section should be performed as appropriate for a particular medical device.

- Concept phase
- Definition phase
- Development phase
- Production phase
- Deployment phase

In the concept phase, past data and future technical projections become the basis for the product/device under consideration. Safety-related problems and their impact are identified and evaluated. The preliminary hazards analysis (PHA) method is a useful tool for identifying hazards during this phase. At the end of this phase, some of the questions to ask with respect to device safety include: (i) are the hazards highlighted and evaluated for the purpose of developing hazard controls? (ii) is the risk analysis started to establish mechanisms for hazard control? and (iii) are the fundamental safety design requirements for the phase in place so that the definition phase can be started?

The purpose of the definition phase is to provide verification of the initial design and engineering associated with the device under consideration. Preliminary hazard analysis is updated along with the initiation of subsystem hazard analysis and their ultimate integration into the device hazard analysis. Techniques such as fault tree analysis and fault hazard analysis may be used to examine specific known hazards and their associated effects. The system definition will initially lead to the acceptance of a desirable general design even though, because of the design's incompleteness, not all associated hazards will be known.

The device development phase includes efforts on areas such as producibility engineering, environmental impact, integrated logistics support, and operational use. Using prototype analysis and testing results, the comprehensive operating hazard analysis is performed to examine man-machine hazards in addition to developing preliminary hazard analysis further because of more completeness of the device's design.

The production phase is also quite critical with respect to safety because the device safety engineering report is prepared using the data obtained in this phase. The report identifies and documents the hazards associated with the resulting device.

During the deployment phase, data relating to failures, accidents, incidents, etc., are obtained, and any changes to the device are carefully reviewed by safety professionals. Safety analysis is updated as appropriate.

6.6 SAFETY ANALYSIS METHODS

Over the years, many system safety analysis methods have been developed, and they can be used to perform safety analysis of medical devices. Some of those methods are as follows:[2,7-11]

- Preliminary hazard analysis (PHA)
- Technique of operations review (TOR)
- Human error analysis (HEA)
- Operating hazard analysis (OHA)

- Fault tree analysis (FTA)
- Failure modes and effects analysis
- Hazard and operability review

The first five of the above methods are described below. The remaining two techniques (failure modes and effects analysis (FMEA) and hazard and operability review (HAZOP)) are discussed in detail in Chapter 7.

6.6.1 Preliminary Hazard Analysis (PHA)

This analysis is normally the first hazard analysis performed on a new product or item; thus, it may be called a conceptual design approach. Furthermore, the approach is relatively unstructured because of the unavailability of information such as definitive functional flow diagrams and drawings. Nonetheless, the term basically means that PHA is performed on the item or device under consideration. Some of the purposes of performing PHA include identifying the safety-critical areas, evaluating hazards, and identifying the safety design criteria to be employed. Usually, PHA is initiated during the primary design phase so that safety aspects are incorporated into trade-off studies.

PHA-related activities include: (i) reviewing relevant past safety experience, (ii) listing of primary energy sources, (iii) investigating energy sources to determine the provisions established for their control, and (iv) determining compliance with safety requirements and other regulations concerning environmental hazards, toxic substances, and personnel safety. Figure 6.3 shows five areas of item/device/system design, for effective hazard identification to be considered by the PHA.[7] The interface safety-related problems include inadvertent activation, electromagnetic interference, and material compatibilities. Examples of the operating, test, maintenance, or other procedural problems are human error and emergency requirements such as egress, rescue, or survival. There are many examples of the energy source hazardous components, but two typical ones are propellants and pressure systems. The normal and abnormal environmental problems include noise, extreme temperatures, X-ray, laser radiation, electrostatic discharge, and shock. The facilities and support equipment with commensurate training for effective usage should be examined carefully with respect to factors such as provisions for storage, assembly, and testing hazardous systems or substances. Table 6.1 presents examples of basic hazards identified by the PHA.[9]

The results of PHA can be used as a guide for performing future detailed analysis. References 7 and 9 present worksheets for conducting PHA.

6.6.2 Technique of Operations Review (TOR)

This method was developed by D.A. Weaver in the early 1970s. It allows both management and employees to work together to conduct analysis of workplace failures, incidents, and accidents.[2] The technique is a hands-on analytical methodology that helps determine the basic item/system causes of an operation failure.[12] It uses a worksheet containing easy-to-understand terms and requiring Yes/No

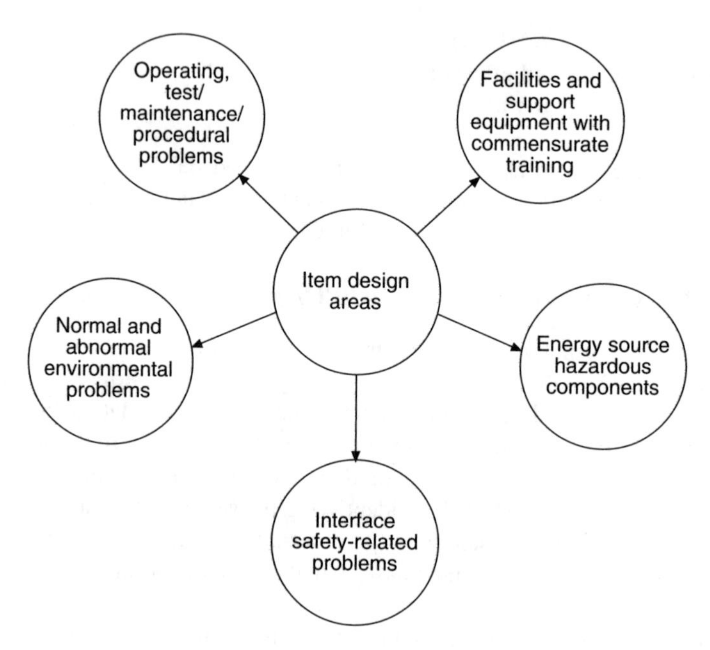

FIGURE 6.3 Areas of item design, for effective hazard identification, to be considered by the PHA.

TABLE 6.1
Some Examples of Basic Hazards Identified by the PHA

No.	Hazard Description
1	Inadvertent release of potential energy.
2	Personnel exposed to radiation over the minimum allowed limit.
3	Electrical shock to personnel.
4	Personnel exposed to extreme heat.
5	Uncontrolled release of corrosive fluid.
6	Uncontrolled cryogenic fluid release.
7	People exposed to high levels of noise.
8	Uncontrolled release of toxic vapor or fluid.
9	Explosion.
10	Inadvertent release of kinetic energy.

decisions. TOR is activated by an incident happening at a certain time and involving certain individuals. The strength of the TOR stems from the involvement of line personnel in the analytical process and its weakness being an after-the-fact process. The TOR method is composed of six steps as shown in Figure 6.4.

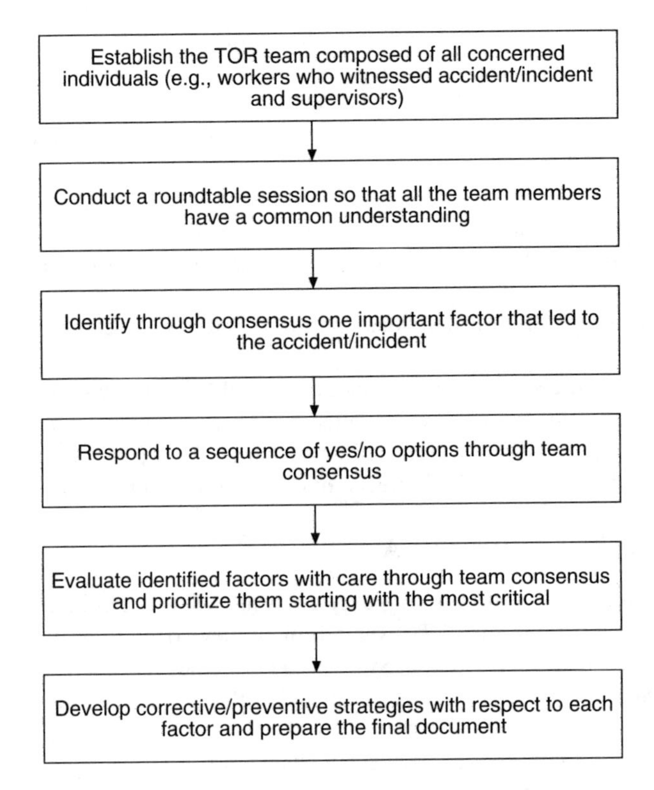

FIGURE 6.4 TOR steps.

6.6.3 Human Error Analysis (HEA)

Human error analysis (HEA) is another important safety analysis method that can be used to identify hazards prior to their occurrence in the form of accidents. There could be two effective approaches to HEA:

- Observing workers during their work hours with respect to hazards
- Carrying out tasks to get first-hand information with respect to hazards

All in all, regardless of the performance of the HEA, it is recommended to conduct it in conjunction with hazard and operability review (HAZOP) and failure modes and effects analysis (FMEA) methods described in Chapter 7.

6.6.4 Operating Hazard Analysis (OHA)

Operating hazard analysis (OHA) is another method used to perform safety analysis, and it focuses on hazards resulting from tasks/activities for operating system functions that happen as the item/system is stored/transported/used. Usually, the OHA is initiated early in the item/system development cycle so that proper input to technical orders is provided, which in turn govern item/system testing.

The application of the OHA will provide a basis for safety considerations such as those listed below:[7,8]

- Special safety procedures with respect to training, handling, transporting, storing, and servicing
- Safety guards and safety devices
- Design modifications to eradicate hazards
- Development of warning, emergency procedures, or special instructions with respect to operation
- Identification of item functions relating to hazardous occurrences

An analyst involved in the performance of the OHA requires engineering descriptions of the item/device/system with available support facilities. OHA is performed using a form requiring information on items such as description of the operational event, hazard description, hazard effects, hazard control, and requirements.

6.6.5 Fault Tree Analysis (FTA)

Fault tree analysis (FTA) is another safety analysis tool that can be used to predict and prevent accidents. In fact, FTA was originally developed to evaluate the safety of the Minuteman Launch Control System.[7] The method is described in detail in Chapter 3. Some of the main points associated with FTA are as follows:

- Extremely useful in the early design phases of new items/devices/ systems
- Very effective in analyzing operational systems/devices/items for undesirable or desirable occurrences
- Permits the user to evaluate alternatives and pass judgement on acceptable trade-offs among them
- Can be used in evaluating certain operational functions, for example, start-up or shut-down phases of item/device/facility operation

Example 6.1

Assume that an overdose of radiation from an X-ray machine may result because of a human error or the machine failure. There are two basic causes for the occurrence of human error: poor operating procedures and unsatisfactory design with respect to human factors. The X-ray machine can fail due to either a hardware failure or a software error. Develop a fault tree for the undesired event: radiation overdose from the X-ray machine.

Figure 6.5 presents a fault tree for the undesired event: radiation overdose from the X-ray machine.

6.6.6 Basic Considerations in Selecting Safety Analysis Methods

For effective applications of safety analysis, methods require a careful consideration in their selection and implementation. Thus, questions on areas, such as those listed

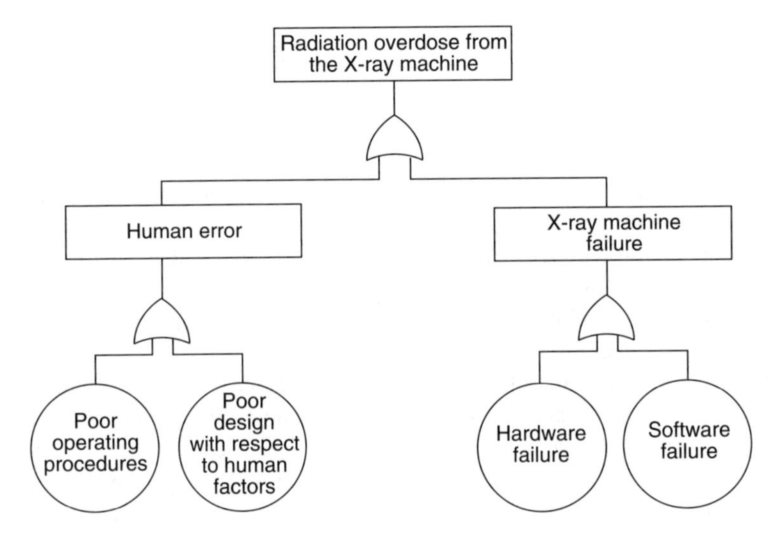

FIGURE 6.5 A fault tree for the top event: radiation overdose from the X-ray machine.

below, should be asked prior to their selection and implementation for a given situation:[8]

- Who will use the end results?
- When are the results required?
- How will you format and detail of the end result data for recipients?
- What type of information, drawings, data, etc. are required prior to the performance of study?
- What is the time frame for the initiation of analysis and its review, completion, submission, and update?
- What mechanism is needed for acquiring information from subcontractors (if applicable)?

6.7 ACCIDENT PROBABILITY ESTIMATION OF A MEDICAL DEVICE

Mathematically, the probability of an accident occurring can be estimated. Thus, the probability of an accident from the operation of a medical device is given by:[13]

$$P_{mda} = X + Y + Z, \tag{6.1}$$

where

$$X \equiv \sum P_{ha}\left(1 + P_{ch} + P_{nh}\right) + P_{hm}\left(1 + P_{ch} + P_{nh}\right)\left(1 - P_{hs}\right) \tag{6.2}$$

$$Y \equiv \sum P_{fa}\left(1 + P_{cf} + P_{nf}\right) + P_{fm}\left(1 + P_{cf} + P_{nf}\right)\left(1 - P_{hs}\right) \tag{6.3}$$

$$Z \equiv \sum (P_{ca} + P_{na})(1 - P_{hs}) \quad (6.4)$$

The symbols used in Equations (6.2)–(6.4) are defined below:

P_{ha} is the probability of occurrence of an irreparable human error that could cause or allow an accident.
P_{ch} is the probability of the device having an adverse characteristic that can lead to human failure.
P_{nh} is the probability of the device encountering an adverse and extraordinary environment that can lead to human failure.
P_{hm} is the probability of occurrence of a repairable human error that can cause or allow an accident. Examples of such human error are a wrong decision, a wrong action, and an unsatisfactory response.
P_{hs} is the probability of correct action exercised as per requirement. For example, correct decision, correct corrective measure, and satisfactory response.
P_{fa} is the probability of occurrence of those failures that will lead to accidents to which no corrective measure is possible.
P_{cf} is the probability of the device having an adverse characteristic that can lead to material failure.
P_{nf} is the probability of the device encountering an adverse, extraordinary environment that can lead to device failure.
P_{fm} is the probability of occurrence of those failures that will lead to accidents unless possible corrective measures are taken in a timely manner.
P_{ca} is the probability of the device having an adverse characteristic that can lead to injury, damage, or loss without the occurrence of material failure or error.
P_{na} is the probability of the device encountering an adverse, extraordinary environment that could lead to injury or damage without the occurrence of failure or error.

6.8 SAFETY COST-ESTIMATION METHODS AND MODELS

Cost is an important factor in safety, and it could determine the success or failure of a medical device in the market, especially with respect to liability, if it causes death or injury. This should be of great concern to manufacturers, as each year many accidental deaths and injuries occur in the U.S. For example, in 1981 there were approximately 99,000 accidental deaths and 9.4 million disabling injuries with a total cost of $87.4 billion.[14] Furthermore, in 1980 employers spent $22 billion to insure or self-insure against job-related injuries and medical costs and compensation payments totaled $3.9 billion and $9.5 billion, respectively.[15]

This section presents cost-estimation methods and models directly or indirectly concerned with medical safety.

6.8.1 Loss-Estimation Model

The loss-estimation model can be used to estimate a medical device manufacturer's loss in dollars. Mathematically, the expected loss or risk is defined by[13]

$$L = \alpha m = \alpha \theta T, \tag{6.5}$$

where

- m is the number of accidents.
- α is the probable loss per accident.
- θ is the predicted or estimated accident rate.
- T is time or other measure during which the accidents happened or could happen.

Example 6.2

Assume that an engineering item has a planned life of 15 years, during which each of the 200 medical devices may be operated annually for 8760 hours. Data on similar items indicate that the losses due to mishaps average approximately $10,000 per accident. The accident occurrence rate is estimated to be 10 times per million hours of operation. It is estimated that the cost of the new item will be twice more than that of the old item. Calculate annual expected loss due to accidents.

Substituting the above given data into Equation (6.5) yields

$$L = (2)(10,000)(10)\left(10^{-6}\right)(200)(8760)$$

$$= \$350,400 \text{ per year}$$

Thus, the annual loss due to accidents will be $350,400.

6.8.2 Heinrich Method

This approach is due to H.W. Heinrich[6] who argued more than 40 years ago that for each dollar of insured cost paid for accidents, there were four times more uninsured costs to the organization. The uninsured costs are also known as indirect costs, hidden costs, invisible costs, etc.

Unfortunately direct or indirect costs of accidents are difficult to state clearly. In fact, they are in the eye of the beholder. Nonetheless, Heinrich defined the indirect cost of an accident as follows:[14,16,17]

$$C_{ia} = CLTM + CLO + CMD + CLTNI + CLTIW + CWM + OCIWNS + CEWS + CCWE + CLTESW + CLP, \tag{6.6}$$

where

C_{ia} is the indirect cost of an accident.
CLTM is the cost of lost time by management.
CLD is the cost of lost orders, etc.
CMD is the cost of machine/material damage.
CLTNI is the cost of lost time on the case by first aid and hospital personnel not compensated by insurance.
CLTIW is the cost of lost time of injured workers.
CWM is the cost due to weakened morale.
OCIWNS is the overhead cost for injured worker while in nonproduction status.
CEWS is the employer cost under welfare and benefit systems.
CCWE is the cost of continuing wages to employees.
CLTESW is the cost of lost time of those employees who stop work to watch or are involved.
CLP is the cost of profit losses on worker's productivity due to idle machines or equipment.

All in all, as some of the above costs are open to speculation, their very existence could be questioned.

6.8.3 Simonds Method

The Simonds method is due to Professor Rollin H. Simonds of Michigan State College, who developed it while working in conjunction with the National Safety Council.[2,18] Professor Simonds' approach is based on the assumption that the cost of an accident can be divided into two categories: insured and uninsured costs. Although insured cost is easy and straightforward to estimate by simply examining accounting records, the estimation of uninsured costs is a challenging task. Nonetheless, Simonds argues that the uninsured cost can be estimated by dividing accidents into four groups and using the following relationship:[2,17,18]

$$C_{ui} = C_1\alpha_1 + C_2\alpha_2 + C_3\alpha_3 + C_4\alpha_4, \tag{6.7}$$

where

C_{ui} is the uninsured cost of an accident.
α_1 is the total number of working days lost because of Class I accidents resulting in temporary total and permanent partial disabilities.
α_2 is the total number of physician cases associated with Class II accidents (i.e., the Occupational Safety and Health Act (OSHA) non-lost workday cases attended by a physician).
α_3 is the total number of first-aid cases associated with Class III accidents (i.e., those accidents in which first-aid was given locally).
α_4 is the total number of noninjury cases associated with Class IV accidents (i.e., those accidents leading to minor injuries but not requiring the services of a medical professional).

C_i is the average uninsured cost associated with the ith Class accidents; for i = 1, 2, 3, and 4.

6.8.4 Medical Device Safety Life Cycle Cost Model

This model can be used to estimate cost of safety for a medical device over its life cycle. Thus, the medical device safety life cycle cost is defined by:[10]

$$MDSC = C_{app} + C_{ins} + C_{rec} + L_{acc} - R_{eim}, \tag{6.8}$$

where

$MDSC$ is the cost of safety for a medical device over its entire life span.
R_{eim} is the reimbursement.
C_{app} is the cost of the accident prevention program.
C_{ins} is the cost of insurance.
C_{rec} is the cost of recalls.
L_{acc} is the accident and claim loss.

6.8.5 Accident Prevention Program Cost-Estimation Model

This model is proposed to estimate cost of the accident prevention program. Thus, mathematically the accident prevention program cost is given by:[16]

$$C_{app} = S_{inp} + C_{ra} + C_{aca} - RE_{eim}, \tag{6.9}$$

where

C_{app} is the cost of the accident prevention program.
RE_{eim} is the reduction in reimbursements.
S_{inp} is the savings in insurance premiums.
C_{ra} is the recall cost that can be avoided.
C_{aca} is the cost of accidents and claims that can be avoided.

6.9 PROBLEMS

1. Describe the following two terms:
   - Medical device reliability
   - Medical device safety
2. What are the types of medical device safety? Discuss them in detail.
3. List at least seven requirements associated with the safe medical device design.
4. Discuss the legal aspects of medical device safety.
5. Discuss safety in the following phases of a medical device:
   - Development phase
   - Production phase
   - Deployment phase
6. Describe the preliminary hazard analysis (PHA).

7. Give at least five examples of basic hazards identified by the PHA.
8. Describe the following two methods of safety analysis:
   - Operating hazard analysis
   - Technique of operations review
9. What are the basic factors to be considered when selecting safety analysis methods?
10. What are the factors proposed by H.W. Heinrich to estimate the indirect cost of an accident?

REFERENCES

1. LaDou, J., Ed., *Introduction to Occupational Health and Safety*, Published by the National Safety Council, Chicago, 1986.
2. Goetsch, D.L., *Occupational Safety and Health*, Prentice-Hall Inc., Englewood Cliffs, NJ, 1996.
3. Fries, R.C., *Reliable Design of Medical Devices*, Marcel Dekker Inc., New York, 1997.
4. Leitgeb, N., *Safety in Electromedical Technology*, Interpharm Press Inc., Buffalo Grove, IL, 1996.
5. Dhillon, B.S., *Reliability in Computer System Design*, Ablex Publishing Corporation, Norwood, NJ, 1987.
6. Bethune, J., Ed., On Product Liability: Stupidity and Waste Abounding, *Med. Dev. Diag. Industry Mag.*, Vol. 18, No. 8, 1996, pp. 8–11.
7. Roland, H.E. and Moriarty, B., *System Safety Engineering and Management*, John Wiley & Sons, Inc., New York, 1983.
8. System Safety Analytical Techniques, Safety Engineering Bulletin, No. 3, May 1971. Available from the Electronic Industries Association, Washington, DC.
9. Gloss, D.S. and Wardle, M.G., Introduction to Safety Engineering, John Wiley & Sons, New York, 1984.
10. Hammer, W., *Product Safety Management and Engineering*, Prentice-Hall, Inc., Englewood, NJ, 1980.
11. Dhillon, B.S., *Advanced Design Concepts for Engineers,* Technomic Publishing Company, Lancaster, PA, 1998.
12. Hallock, R.G., Technic of Operations Review Analysis: Determine Cause of Accident/Incident, *Safety and Health*, Vol. 60, No. 8, 1991, pp. 38, 39, and 46.
13. Hammer, W., *Product Safety Management and Engineering*, Prentice-Hall, Inc., Englewood Cliffs, NJ, 1980.
14. Ferry, T.S., *Safety Program Administration for Engineers and Managers*, Charles C. Thomas Publisher, Springfield, IL, 1984.
15. Lancianese, F., The Soaring Costs of Industrial Accidents, *Occupational Hazards*, August 1983, pp. 30–35.
16. Heinrich, H.W., *Industrial Accident Prevention: A Scientific Approach*, McGraw-Hill, New York, 1959.
17. Raouf, A. and Dhillon, B.S., *Safety Assessment: A Quantitative Approach*, Lewis Publishers, Boca Raton, FL, 1994.
18. National Safety Council, Accident Prevention Manual for Industrial Operations: Administration and Programs, Chicago, 1988.

7 Medical Device Risk Assessment and Control

CONTENTS

7.1 INTRODUCTION

The history of risk in the medical field may be traced back to the 4th century B.C. when Hippocrates, an ancient Greek doctor, correlated occurrences of diseases with environmental exposures.[1,2] In the 16th century A.D., Agricola established the correlation between occupational exposure to mining and health. Also, physicians in the Middle Ages became aware of a correlation between chemicals or agents and health. For example, John Evelyn (1620–1706) concluded that smoke was the cause of respiratory problems in London.

The roots of risk analysis extend to the works of Pierre Simon de Laplace. In 1792, he calculated the probability of death with and without smallpox vaccination. In the 20th century, the conceptual developments of risk analysis in the U.S. were due to factors such as the development of nuclear power plants and concerns about their safety, as well as the establishment of various government agencies or bodies, such as, the Occupational Safety and Health Administration (OSHA) and Environmental Protection Agency (EPA). Additional, information on the history of risk analysis and risk management can be found in reference 1. Risk may be expressed as a measure of the probability and security of a negative effect to health, environment, equipment, or property. The entire process of risk assessment and risk control is known as risk management. In medical devices, risk assessment and control is important from various perspectives, including safety and economics.

This chapter presents different aspects of risk management, assessment, and control directly or indirectly associated with medical devices.

7.2 RISK MANAGEMENT, DEFINITION AND PROGRAM STEPS

One of the most widely used definitions of the risk management is that it is the systematic application of management practices, policy, and procedures to identifying, analyzing, controlling, and monitoring risk.[3] The main objective of risk management is the efficient planning of resources required to obtain financial balance and operating effectiveness subsequent to a fortuitous loss, thus achieving a short-term cost of risk stability and long-term risk minimization,[4]

Basically, a risk management program involves the following steps:[5]

- Developing definitions of requirements and mechanisms to achieve them
- Outlining associated responsibilities and accountability
- Identifying what requires authorization and responsibility for handling it
- Defining the knowledge and skills required for the system implementation and a provision for training those without the needed skills
- Developing documentation to demonstrate conformance to specified policies and procedures
- Establishing and incorporating appropriate measures to cross-check and verify that procedures are being followed effectively
- Verifying that the systems are in place and operating effectively

7.3 RISK ASSESSMENT

Risk assessment is the process of risk analysis and risk evaluation. Risk analysis is composed of steps such as scope definition, hazard identification and risk determination, and it uses available data to determine risk to humans, environment, property, or equipment from hazards. Risk evaluation is the stage at which values and judgements make entry to the decision process.

7.3.1 Risk Analysis Process and Risk Analysis in Item Life Cycle Phases

The risk analysis process is made up of the following six steps:

- Development of scope definition
- Identification of hazards
- Estimation of risk

- Documentation
- Verification of results
- Periodic update of analysis

The development of scope definition can be accomplished in five steps: (i) describing the problems leading to risk analysis and then developing risk analysis objectives on the basis of important identified concerns; (ii) defining the system under study by considering factors such as definition of functional and physical boundaries, definition of environment, and description of the general systems; (iii) identifying assumptions and constraints associated with risk analysis; (iv) highlighting the decisions to be made; (v) documenting the overall plan. The identification of hazards step is concerned with highlighting the hazards that will lead to risk in the system. More specifically, it calls for the preliminary evaluation of the significance of the highlighted hazardous sources. The main objective of this evaluation is to uncover the effective course of action.

Risk estimation is carried out in seven steps: (i) investigate hazard sources for determining the likelihood of occurrence of the originating hazard and its associated consequences; (ii) perform pathway analysis for determining mechanisms and the likelihood through which the receptor under study is influenced; (iii) select risk estimation method; (iv) define data requirements; (v) describe assumptions and other related matters associated with methods and data being used; (vi) estimate risk to determine the degree of influence of the receptor under study; (vii) document the study.

The fourth step, documentation, involves effectively documenting three items: the risk analysis plan, the preliminary evaluation, and the risk estimation. A typical documentation report contains items such as those listed below:

- Title
- Abstract
- Objectives/scope
- Table of contents
- Assumptions/limitations
- System description
- Description of analysis methodology
- Hazard identification results
- Description of the model and associated assumptions
- Quantitative data and associated assumptions
- Risk estimation results
- Discussion of results
- Sensitivity analysis
- Conclusions
- References
- Appendices

Verification of results is a review process that determines the integrity and accuracy of the risk analysis process. The verification is performed at appropriate times and levels by person(s) other than the involved individual(s).

The last and final step concerns the periodic update of analysis as new information becomes available.

Risk analysis is associated with a hazardous system's entire life cycle as well as one specific phase of it. The complete life cycle of the system may be divided into three categories: (i) the concept and definition, design and development phase; (ii) the construction and production, operation and maintenance phase; (iii) the disposal phase.

Some of the risk analysis objectives belonging to the first phase are as follows:

- Identifying major contributors to risk
- Providing input to establish procedures for normal and emergency conditions
- Providing input to the design process
- Assessing overall design adequacy
- Providing input to the evaluation of the acceptability of proposed potentially hazardous facilities or activities

The risk analysis objectives belonging to the second phase include the following:

- Gauging and evaluating experience for comparing actual performance and relevant requirements
- Updating information on major risk contributors
- Providing appropriate inputs to the optimization of normal and emergency procedures
- Providing input on plant risk status in operational decision-making

The category three phase objectives include assessing the risk associated with process disposal activities with respect to satisfying specified requirements effectively, and providing input to disposal policies and procedures.

7.3.2 Risk Analysis Methods

There are many methods available to perform risk analysis of engineering systems,[6-7] and they can be applied to medical devices. As each method has its strong and weak points, a careful consideration is necessary prior to selecting one. Factors such as those listed below should be considered when deciding their applications:

- Objective of the study
- Appropriateness to the system under consideration
- Simplicity to use
- Scientific defensibility
- Result format with respect to improvement in understanding the risk occurrence and risk controllability
- Requirement for manpower and other resources
- Level of expertise required
- Requirement for information and data

- Level of risk
- Updating flexibility

The risk analysis approaches used to analyze engineering systems may be classified into two groups: hazard identification and risk estimation.[6]

Hazard Identification Methods

Three important methods of hazard identification are failure modes and effect analysis (FMEA), hazard and operability study (HAZOP), and event tree analysis (ETA). Each method is described below.

Method I: Failure Modes and Effect Analysis (FMEA)

This is an important tool of risk analysis. It was developed in the early 1950s to evaluate safety of flight control systems.[10] The method is described in detail in Chapter 3. Seven basic steps are involved in performing failure modes and effect analysis:

- **Establishing system definition**. This involves defining system boundaries and detailed requirements.
- **Listing components and subsystems**. This means listing all components and subsystems associated with the system under consideration.
- **Listing component failure modes**. This involves listing all relevant failure modes of components associated with the system as well as description and identification of the components.
- **Assigning failure rates**. This involves assigning failure rates to each failure mode of system components.
- **Identifying failure mode effect or effects**. This involves listing effect(s) of each failure mode on subsystem and plant.
- **Entering remarks**. This involves entering remarks for each failure mode.
- **Reviewing critical failure modes**. This means reviewing each critical failure mode and taking appropriate measures.

In real life, FMEA is performed by filling required information in FMEA worksheets.

Method II: Hazard and Operability Study (HAZOP)

This technique was originally developed for application in chemical industries and is basically a form of FMEA. In the past, HAZOP has proven to be an effective tool for uncovering hazards designed into facilities or introduced into existing ones due to actions such as changes made to process conditions or operating procedures.

The approach has the following three-fold objectives:

- Develop a full description of the facility under consideration.
- Review each facility component to highlight how deviations from the design intentions can happen.
- Determine if these deviations can lead to hazards or operating problems.

Although, HAZOP can be used at various stages of design and carry out analysis of process plants in operation, its application during the early stages of design often result in a safer detailed design. HAZOP is composed of five basic steps: (i) establish objectives and scope of the study; (ii) form a HAZOP team by including members from design and operation; (iii) obtain appropriate drawings, process description, and other related information (for example, layout drawings, process flow sheets, and operations and maintenance procedures); (iv) analyze all major pieces of equipment, supporting equipment, and so on; (v) document the study in an effective manner.

Method III: Event Tree Analysis (ETA)

This technique is often used to analyze systems more complex than those handled by the FMEA method. It is a "bottom-up" approach that identifies the possible outcomes when the occurrence of the initiating event is known.[9,11-12] In the past, ETA was found to be quite effective in analyzing facilities that possess engineered accident-mitigating characteristics to identify sequence of events subsequent to the initiating event and generate given sequences. In this method, it is usually assumed that the outcome of each sequence event is either a success or a failure.

More specifically, it may be said that the ETA method identifies the relationship between the success or failure of various mitigating systems and the hazardous event subsequent to the initiating event. Because of the inductive nature of the ETA approach, the basic question asked is "what happens if…?" Some of the additional factors associated with the method are excellent tools to identify events that need further investigation using fault tree analysis (FTA). It is difficult to incorporate delayed success or recovery events when conducting ETA, and it leaves room for missing some important initiating events.

Risk Estimation Methods

Two well-known methods of risk estimation are consequence analysis and frequency analysis. Both methods are described below.

Method I: Consequence Analysis

This is an important risk estimation method, and it is used to estimate the impact of the undesired event on adjacent people, environment, or property. For risk estimation with respect to safety, it involves computing the probability that people at varying distances and varying environments from the source of the undesired event will suffer injury/illness. Typical examples of an undesired event include fires, explosions, and release of toxic materials. For the effectiveness of the study, it is important that consequence analysis takes into account factors such as the following:

- Basis for analysis with respect to selected undesirable events
- Measures for eradicating consequences
- Explanation of the criteria employed for accomplishing the identification of consequences

- Immediate and aftermath consequences
- Description of any series of consequences resulting from the undesirable events under consideration

Method II: Frequency Analysis

This method estimates each undesired event's or accident scenario's frequency of occurrence. Often, the following two approaches are used to perform frequency analysis:

- Using the past frequency data of undesired events to predict frequency of their occurrences in the future
- Employing techniques such as FTA and ETA to determine occurrence frequency of undesired events

As each of the above two approaches has strengths and weaknesses, whenever possible each approach should be used to serve as a check for the other.

7.3.3 Risk Analysis Benefits

There are many advantages associated with the performance of risk analysis. The important ones are shown in Figure 7.1.

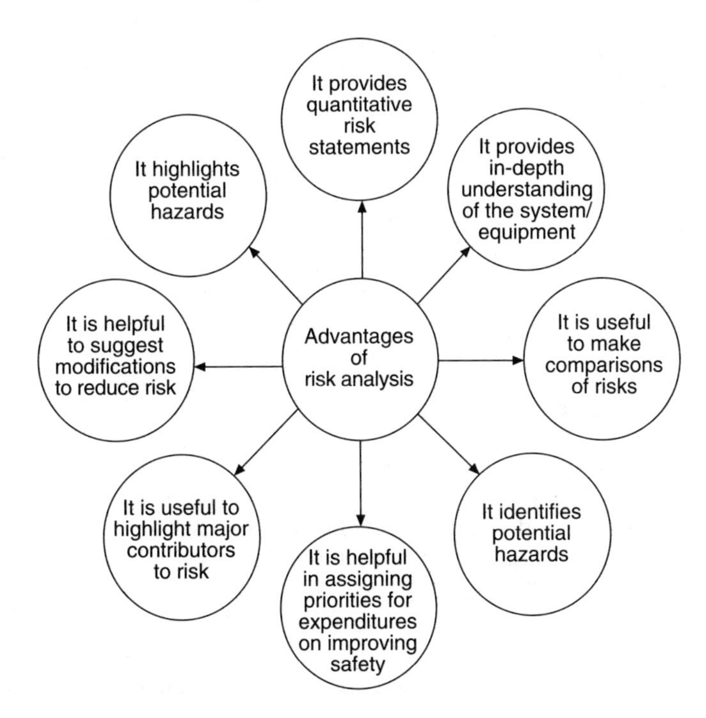

FIGURE 7.1 Advantages of risk analysis.

7.4 MEDICAL DEVICE RISK ASSESSMENT

Medical device manufacturers who produce a wide range of products for use by health care professionals or patients must deal with various government regulatory agencies whose responsibility is to enforce factors such as safety, proper record keeping, and effectiveness. One example of such agency is the Food and Drug Administration (FDA) in the U.S. In one form or another, the concept of risk assessment appears in various decision-making by an agency such as the FDA.[13]

7.4.1 CRITICAL FACTORS IN MEDICAL DEVICE RISK ASSESSMENT

Past experience indicates that factors that are critical to medical device risk assessment relate both to the device and its usage. Figure 7.2 presents critical factors related to both of these areas.[13]

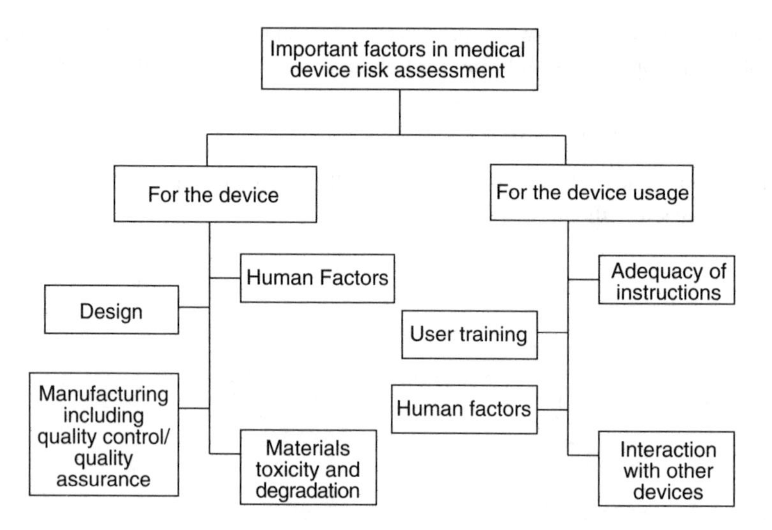

FIGURE 7.2 Critical factors in medical device risk assessment.

Design and Manufacturing

A medical device could be hazardous if inadequate attention is given to design elements that influence performance, or if the device is poorly manufactured. For example, an oversight to design a structural part that resists the expected stress may result in its fracture. During the manufacture of the device, good manufacturing practices do not eradicate inherently bad designs and may simply help produce a "bad design."

On the other hand, improper attention to a manufacturing quality assurance program may cause a device defect. For example, inappropriate stressing of a part during a manufacturing process may weaken that part which subsequently, during its application, may end up fracturing.

In both of the above cases — i.e., faulty design and flawed design implementation — the type of hazard may not be that different. It is important however, that risk assessment makes a distinction between the two or determines if both are involved at a reasonable level. Knowledge of this type of information is essential because the approaches to risk management in each case will be different.

Device hazard is considered with regard to factors such as the device causing harm, and the device failing to provide its intended benefit. Two examples of these conditions are

- An electric shock from a supposedly well-insulated device can harm individuals such as health care professionals and patients.
- An error in diagnosis could result in incorrect therapy, or no therapy at all, causing severe consequences for the patient involved.

All in all, under conditions such as the above, it is important to the eventual management of risk to evaluate if the involved hazards are caused by omission or commission.

Human Factors

These are design elements and use conditions associated with a medical device that significantly influence interactions between the device and the user. An additional complexity is placed by the interactions on the ability to highlight hazards that might be concerned with a specific device. Poor attention given to human factors in early medical devices resulted in various types of problems. It may be added that without considering human factors as components of risk assessment, it would be extremely difficult to consider the proper risk management approach for medical devices.

Materials Toxicity and Degradation

Appropriate material selection is an element of a good design. Toxicity is a key element in determining the risk of a material used in a device for implant. It can be difficult to determine the material used in a device because a device can contain the following:

- Contaminant material
- Catalytic material used in synthesis
- Residual material to which the device became exposed during its manufacturing processes
- The material making up the primary molecular structure

Factors such as those above lead to questions such as: (i) what amount of toxic agent is present in the device? (ii) which one is the toxic agent and does it leach from the device? and (iii) is the body part exposed sensitive to injury due to the toxic agent? The tissue environment could be an influencing factor for leachability, and it can affect whether the toxic agent is distributed throughout the system or remains locally. Furthermore, the tissue environment can be quite hostile and can lead to initiating, promoting, or accelerating the material degradation. Generally, the hostility of a particular environment depends on two factors: material type and the amount of time it is in the given environment. Risk assessment has to consider factors such as these as well as an associated judgement.

Device Users

Past experience indicates that the way a medical device is used plays a significant role in its overall safety and effectiveness. Examples of device users are health care professionals, patients, and family members of patients. Also, there could be a large variation in users' skills with the devices due to various factors, including health status, motivation, education and training, and environment. Furthermore, users represent the decision point on the usage of medical technology. Decisions could be taken incorrectly. One example is the improper use of oxygen therapy for premature infants by the health care professionals, which is proven to damage the eye retina.

Although the types of hazards that may result from user error are not unusual, the incidence of occurrence of such hazards may vary from nonexistent to frequent depending on user influence. Thus, in assessing the risks such as the ones discussed above, factors such as the degree of user training and the effectiveness of the instruction provided with the device are directly related.

Interaction with Other Devices

Under certain circumstances many medical devices may be used within close proximity to one another. A scenario such as that may have been considered in the design, but the worst case design allowance could be bypassed by the ingenuity of the device users. For example, electromagnetic interference from one device to another could be hazardous to patients.

Another example of interaction with other devices is that the users may try to interchange incompatible device parts during the repair process, thus causing a risk to the patient. Risk assessment must consider factors such as those listed above which concern the interaction of devices.

7.4.2 Case Studies of Risk Assessment in Medical Devices

Although there are many real-life examples of medical devices with respect to risk assessment directly or indirectly in the documented literature, in this section we present two such examples concerning heart valves and anesthesia systems. Both are described below.[13]

Heart Valves

This example concerns a prosthetic heart valve introduced in 1979. This new development was considered a major step forward with respect to reducing the frequency of clot formation around the valve. After some time in the market it was observed that the valve malfunctioned due to a rare but persistent pattern of mechanical failure. However, a close analysis of data did not exhibit the pattern of mechanical failure. Nonetheless, the analysis of the manufacturing process indicated several changes in procedure to eradicate repetitive bending of an individual component that could be weakened by the bending.

Similarly, the examination of the design specifications revealed that a specific valve closure mode could also exist that could lead to irregular stress placed on exactly the same component. Under such condition, if the problem was the faulty manufacturing process, then the changes to the process could have been sufficient to eradicate the mechanical failure. However, on the other hand, if there was a flaw in the design, the process changes would not necessarily eliminate the problem in question. Therefore, in the final risk assessment, it was impossible to distinguish between the design and the manufacturing process or both as the basic cause of the persisting problem.

Subsequently, due to an unacceptable rate of mechanical failures, the manufacturer of the valve removed its product from the market.

Anesthesia Systems

Past data revealed that surgery patients being given anesthesia gas and being supported by mechanical ventilation could very well be disconnected from the system due to an accident. After performing a careful analysis of this risk, it was concluded that the equipment generally was well-manufactured, but there was room to improve design of the connectors and a mechanism could be added to indicate when a disconnect happens.

Furthermore, an associated risk assessment investigation concerning user behavior revealed that anesthesia incidents occur due to user error in setting up the equipment and failure to detect it. To overcome these problems, manufacturers, the FDA, and users in the U.S. have joined forces to provide a checklist procedure with characteristics such as the following:

- It can be quickly run through.
- It detects most preexisting problems.
- It assures appropriate attention related to setup issues — including the need to avoid disconnects.

All in all, it may be added that risk assessment that overlooks either the user or device role may result in ineffective risk management.

7.5 INTEGRATING RISK ASSESSMENT WITH MEDICAL DEVICE DESIGN CONTROL

A study of voluntary device recalls for 1983–1989 revealed that 44% were attributed to errors or deficiencies concerning device design that could have been prevented by effective design controls.[3] Risk management performed effectively as an element of a design control program could prove to be an effective tool in reducing the problems resulting in recalls. Also, the FDA's quality system regulation concerning medical devices requires that "design validation shall include… risk analysis."[3] More specifically, the FDA expects risk analysis to be integrated with the device design control.[14]

Risk assessment can be integrated with design control activities shown in Figure 7.3.[3]

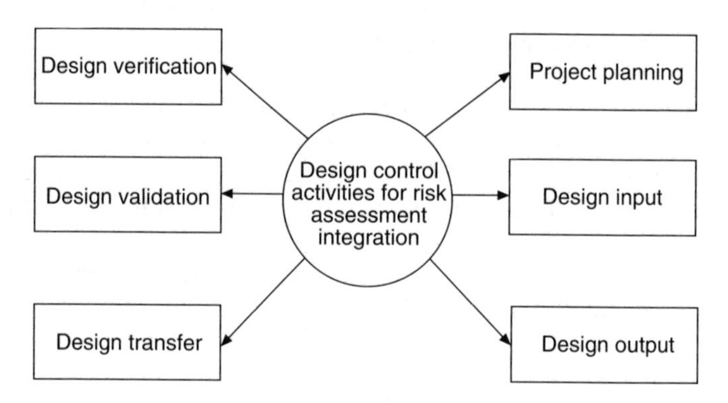

FIGURE 7.3 Design control activities for risk assessment integration.

The activities shown in Figure 7.3 are discussed below with respect to risk assessment:

- **Project planning**. The risk management plan should be a part of the product development or project management plan, and it should incorporate a strategic approach for identifying and controlling risk in the product development life cycle.
- **Design input**. In this case, the essential design inputs are the existing safety standards and safety requirements highlighted in risk assessments.
- **Design output**. The important design outputs are the risk reduction measures introduced into product design.
- **Design verification**. This affirms that all relevant safety requirements are taken into consideration by the risk reduction measures in the device design.
- **Design validation**. This should demonstrate that all relevant safety requirements can be effectively and consistently satisfied by considering user/patient needs and intended application.
- **Design transfer**. Each formal design review should address risk assessment activities, as required, to make sure that specified actions are assigned/monitored before the design transfer takes place.

All in all, the risk management activities do not end with design transfer. They should also be integrated with production and distribution.

7.6 FACTORS AFFECTING THE VALUES OF CONSEQUENCES AND MEDICAL DEVICE RISK ASSESSMENT-RELATED DATA

There are many factors that affect the value of consequences. Such factors can be grouped into three categories:[15]

- **Group I**: This group contains factors that involve types of consequence.
- **Group II**: This group contains factors that involve the nature of consequences.
- **Group III**: This group contains miscellaneous factors.

The factors in Group I can be divided into four areas: spatial distribution and discounting of risks (such as identification of risk agents, spreading risk, and geographic distribution of risk), controllability of risk (such as systemic control of risk, crisis management, and perceived degree of control), voluntary and involuntary risks (such as degree of knowledge, availability and alternatives, equity and inequity, and imposition-exogenous and endogenous), and discounting in time.

The factors in Group II include natural vs. man-originated events, hierarchy of need fulfillment, common vs. catastrophic risks, variation in cultural values, knowledge as a risk, and national defense.

The factors in Group III may be divided into four areas: propensity for risk taking (such as conflict avoidance, group and individual), situational factors (lifesaving systems and surprise and dissonant behavior), and factors involving the magnitude of a consequence occurrence frequency (such as spatial distribution of risks and high frequency levels, and low frequency levels and threshold).

Over the years, professionals working in the risk assessment area have come up with estimates concerning various causes and events. Tables 7.1 and 7.2 present such estimates that are directly or indirectly related to medical devices.[16,17] Data concerning the incidence of adverse events and negligence in hospitalized patients are given in reference 18.

TABLE 7.1
Average Loss of Life Expectancy Due to Selective Causes[16,17]

Item No.	Cause	Average Life Expectancy Loss in Days
1	Electrocution	5
2	Medical X-rays	6
3	Legal drug misuse	90
4	Illicit drugs	18
5	Safest jobs — accidents	30
6	All catastrophes combined	35

TABLE 7.2
Annual Probability Estimates for the Occurrence of Selective Events[16]

Item No.	Event Description	Probability Estimate
1	Drug toxicity	10^{-4} to 10^{-6}
2	Chemical explosion and fire	10^{-5} to 10^{-7} per site
3	Aircraft crash (for a selected site adjacent to a selected airport)	10^{-5} to 10^{-7}
4	Fire	10^{-6} per high risk site
5	Earthquake	10^{-3} to 10^{-4}
6	Train crash	10^{-1} to 10^{-2}

7.7 PROBLEMS

1. Write an essay on the history of risk analysis.
2. Define the following terms:
   - Risk management
   - Risk control
   - Risk assessment
3. Discuss the risk analysis process.
4. Describe the risk analysis method known as hazard and operability study (HAZOP).
5. Compare the following two methods:
   - Event tree analysis
   - Consequence analysis
6. What are the advantages of performing risk analysis?
7. What are the critical factors in medical device risk assessment?
8. Describe some case studies of risk assessment in medical devices.
9. What are the design control activities associated with the risk assessment integration?
10. What are the factors that affect the values of consequences?

REFERENCES

1. Covello, V.T. and Mumpower, J., Risk Analysis and Risk Management: A Historical Perspective, *Risk Analysis*, Vol. 5, 1985, pp. 103–120.
2. Molak, V., Ed., *Fundamentals of Risk Analysis and Risk Management*, CRC Press, Boca Raton, FL, 1997.
3. Knepell, P.L., Integrating Risk Management with Design Control, *Med. Dev. Diag. Ind. Mag.*, October 1998, pp. 83–91.
4. Greene, M.R. and Serbein, O.N., *Risk Management: Text and Cases*, Reston Publishing Company, Reston, VA, 1983.

5. Ozog, H., Risk Management in Medical Device Design, *Med. Dev. Diag. Ind. Mag.*, October 1997, pp. 112–120.
6. CAN/CSA-Q3640-91, Risk Analysis Requirements and Guidelines, prepared by the Canadian Standards Association, 1991. Available from the Canadian Standards Association (CSA), 178 Rexdale Boulevard, Rexdale, Ontario, Canada.
7. Covello, V. and Merkhofer, M., Risk Assessment and Risk Assessment Methods: The State-of- the-Art, National Science Foundation (NSF) report, 1984, NSF, Washington, DC.
8. Dhillon, B.S. and Rayapati, S.N., Chemical Systems Reliability: A Survey, *IEEE Transactions on Reliability*, Vol. 37, 1988, pp. 199–208.
9. Wesely, W.E., *Engineering Risk Analysis, in Technological Risk Assessment,* Rice, P.F., Sagan, L.A., and Whipple, C.G., Eds. Martinus Nijhoff Publishers, The Hague, Netherlands, 1984, pp. 49–84.
10. Countinho, J.S., Failure Effect Analysis, *Transactions of the New York Academy of Sciences*, Vol. 26, 1964, pp. 564–584.
11. Ramakumar, R., *Engineering Reliability: Fundamentals and Applications*, Prentice-Hall, Englewood Cliffs, NJ, 1993.
12. Cox, S.J. and Tait, N.R.S., *Reliability, Safety, Risk Management*, Butterworth-Heinemann Ltd., Oxford, UK, 1991.
13. Anderson, F.A., Medical Device Risk Assessment, *The Medical Device Industry*, Estrin, N.F., Ed., Marcel Dekker Inc., New York, 1990, pp. 487–493.
14. Design Control Guidance for Medical Device Manufacturers, Food and Drug Administration (FDA), Rockville, MD, March 1997.
15. Rowe, W.D., *An Anatomy of Risk*, John Wiley & Sons Inc., New York, 1977.
16. McCormick, N.J., *Reliability and Risk Analysis*, Academic Press, Inc., New York, 1981.
17. Cohen, B.L. and Lee, I.S., *Health Phys.*, Vol. 36, 1979, p. 707.
18. Brennan, T.A., et al., Incidence of Adverse Events and Negligence in Hospitalized Patients, *N. E. J. Med.*, Vol. 324, 1991, pp. 370–376.

8 Medical Device Quality Assurance

CONTENTS

8.1 INTRODUCTION

Because quality is important in the manufacture of medical devices, manufacturers are under increasing pressure to follow more closely the quality systems to ensure that the manufactured items are effective, safe, reliable, and that they meet applicable specifications and standards.

The history of quality assurance may be traced back to ancient times. For example, around 1450 B.C. Egyptian wall paintings show evidence of inspection-related activities.[1] However, in modern times it was only in the early 1900s when an inspection department was set up by the Western Electric Company.[2] Subsequently, the department was renamed the Quality Assurance Department, and in 1924 Walter A. Shewhart (at this time Western Electric was a subsidiary of Bell Telephone Laboratories) developed control charts to determine whether a given manufacturing process was within or out of control. In 1946, the American Society for Quality Control was formed, and since that time or earlier, many other new developments in the field have taken place. A detailed history of quality assurance is given in reference 3. Amendments to the Federal, Food, Drug, and Cosmetic Act of 1976 concerning medical devices have established a complex statutory framework to allow the The Food and Drug Administration (FDA) to regulate almost all aspects of medical devices, from testing to marketing, thus putting more pressure on the quality assurance programs of medical devices.

This chapter presents various aspects of quality assurance in medical devices.

8.2 TERMS AND DEFINITIONS

In the quality assurance field, many terms and definitions are used. This section presents some of the terms and definitions useful for assuring the quality of medical devices.[4-6]

- **Quality**. This is the totality of features or characteristics of an item that influence its ability to meet requirements of users.
- **Quality Assurance**. This is a planned and orderly sequence of all actions appropriate to provide satisfactory confidence that the product conforms to specified technical requirements.
- **Inspection**. This is the process of examining, measuring, testing, or otherwise making comparisons of the item with the specified requirements.
- **Quality Management**. This is the totality of functions concerned with the evaluation and achievement of quality.
- **Quality System**. This is the organizational set up, responsibilities, resources, processes, etc. for implementing quality management.

- **Quality Measure**. This is a quantitative measure of the characteristics and features of an item or service.
- **Relative Quality**. This is the degree of excellence of an item or service.
- **Quality Policy**. This is the management's set goals and approach in achieving quality.
- **Total Quality**. This is a business philosophy that involves all individuals for continuously improving an organization's performance with respect to quality.
- **Quality Function Deployment**. This is a systematic and structured technique to highlight customer requirements, translate them into a realizable item or service parameters, and guide or direct the process of implementation in a manner that brings about a competitive advantage.

8.3 REGULATORY COMPLIANCE OF MEDICAL DEVICE QUALITY ASSURANCE

To produce better quality medical devices, agencies such as the FDA and the International Organization for Standardization (ISO) are playing an important role through their good manufacturing practices (GMP) regulation and ISO 9000 family of quality system standards, respectively. Both demand a comprehensive approach to medical device quality from manufacturers. The GMP regulation went into effect June 1, 1997, but the FDA granted a grace period to the manufacturer until June 14, 1998. During this time, the FDA could have inspected a manufacturer's facilities with respect to the Quality System Regulation, but it would not have listed any deficiencies on Form 483, nor would it have brought any sanctions against manufacturers for noncompliances.

A detailed description of the ISO 9000 requirements is given in reference 5. A mechanism to meet both the GMP regulation and applicable ISO 9000 requirements is described below.

8.3.1 Procedure to Satisfy GMP Regulation and ISO 9000 Requirements with Respect to Quality Assurance

This procedure is published in the form of a quality assurance manual, and assists manufacturers of medical devices to comply with regulatory requirements in a straightforward, organized, and effectively documented manner.[7] The approach is divided into three areas:

- Area A
- Area B
- Area C

Area A outlines the company policy with respect to the manufacture of medical devices and the authority and responsibility of the quality assurance department to implement the policy, and defines the type of records to be maintained.

Area B involves defining the company policy in regard to the administration of the quality assurance department and its subdivision such as receiving inspection, standards laboratory, and tool and gage inspection. It is also concerned with outlining policy for internal audits, product qualification, and the organizational chart.

Area C concerns outlining quality assurance directives useful for implementing and monitoring device conformance and procedural compliance according to the GMP regulation. (These directives also apply to the ISO 9000 requirements).

These directives cover 19 distinct sub-areas: quality audits, sterilization process and control, design control, equipment, statistical quality control and sampling, procedure for FDA inspection, facility control for the manufacturing of medical devices, complaint report procedure, control of measuring equipment and tooling, control of inspection stamps, field corrective action, lot numbering and traceability, supplier and subcontractor quality audits, failure analysis, receiving inspection, personnel, in-process and final inspection, component/warehouse control, approval and control of labels, labelling, and advertisement.[7]

8.4 MEDICAL DEVICE DESIGN QUALITY ASSURANCE PROGRAM

As in the case of any other engineering product, the design phase is the most important phase in the medical device's life cycle. It is the phase when the device's inherent reliability, safety, and effectiveness are established. Furthermore, it may be said that regardless of the degree of carefulness exercised during the manufacture or the effectiveness of the GMP program, the inherent safety and effectiveness of a medical device cannot be improved except through design enhancement. The design quality assurance program is a key factor in this regard. The FDA has played a leading role in getting manufacturers to develop design quality assurance programs by publishing a document entitled "Preproduction Quality Assurance Planning: Recommendations for Medical Device Manufacturers."[8] This document outlines useful design practices applicable to medical devices, thus assisting manufacturers in planning and implementing their preproduction quality assurance programs.

There are 12 elements in the preproduction or design quality assurance program recommended by the FDA: (i) organization, (ii) specifications, (iii) design review, (iv) reliability assessment, (v) parts and materials quality assurance, (vi) software quality assurance, (vii) labelling, (viii) design transfer, (ix) certification, (x) test instrumentation, (xi) personnel, and (xii) quality monitoring after the design phase.

8.4.1 Organization

This is concerned with the organizational aspects of the preproduction or design quality assurance program, for example, the organizational elements and authorities appropriate to develop the program, to execute program requirements, formal documentation of the specified program goals, formal establishment of audit program, etc.

8.4.2 Specifications

After establishing physical, performance, and chemical characteristics for the proposed device, the characteristics should be translated into formally documented design specifications through which the design can be developed, controlled, and evaluated. These specifications should address factors such as safety, reliability, precision, and stability. Furthermore, in establishing physical configuration, performance, safety, and effectiveness goals of the design, factors such as the user environment, the user, and the expected use of the device should be considered.

It is also important for professionals belonging to areas such as quality assurance, manufacturing, marketing, research and development, and reliability to review and evaluate the specification document. Careful consideration should be given to the following two factors in the specification:

- **System compatibility**. This involves the device's compatibility with other devices in the intended operating system to assure proper functioning of the overall system. For example, make sure you include breathing circuits with ventilators and disposable electrodes with cardiac monitors.
- **Design changes**. These are changes made to the specification during research and development that are accepted as design changes. Such changes must be well-documented and carefully reviewed so that they do not compromise safety or effectiveness and that they effectively accomplish the intended goals.

8.4.3 Design Review

The purpose of design review is to identify and rectify design deficiencies as early as possible because they will be less costly to implement. The design review program should be well-documented and include items such as organizational units, process flow diagrams, procedures, schedule, and a checklist of variables. Although the extent and frequency of design reviews will depend on the complexity and significance of the device under study, the assessment should include the following items:

- Subsystems
- Software (if applicable)
- Components
- Labelling
- Packaging
- Support documentation (i.e., drawings, instructions, test specifications, etc.)

The members of the design review team should be from areas such as manufacturing, research and development, quality assurance, servicing, marketing, engineering, and purchasing. Also, when considered appropriate, design reviews should include the performance of failure mode and effect analysis, i.e., failure mode, effects and criticality analysis (FMECA) and fault tree analysis (FTA). Both methods are described in Chapter 3.

8.4.4 Reliability Assessment

This may be described as the process of prediction and demonstration used to estimate the basic reliability of an item or device. Reliability assessment should be carried out for new and modified designs, and its appropriateness and extent should be determined by the degree of risk the device presents to the user. Reliability assessment is started by theoretical and statistical approaches by first determining the reliability of each element/part/component and ultimately the entire device or system. Remember that this approach provides only an estimate of reliability. For better or proper assessment, the device or system should be tested under a simulated use environment. However, the most meaningful reliability data can be obtained only from actual field use.

All in all, reliability assessment is an essential element of the preproduction quality assurance program and is a continual process that includes reliability prediction, demonstration, and data analysis, then reprediction, redemonstration, and data reanalysis on a continual basis.

8.4.5 Parts and Materials Quality Assurance

This is concerned with assuring that parts and materials used in device or product designs have an appropriate level of reliability to achieve their set goals. This requires establishment and implementation of comprehensive parts and materials quality assurance programs by the medical device manufacturers. These programs should encompass areas such as selection, specification, and qualification and ongoing verification of the quality of parts and materials, and whether the parts are purchased from vendors or fabricated in-house.

Parts and materials should be categorized according to the severity of their effect on safety, reliability, and effectiveness in the event of their failure to achieve the set goal. And their acceptability for chosen applications should be evaluated and supported by both observed and computed test data. All in all, failure of parts and materials during qualification to satisfy specified safety, performance, and effectiveness goals should be thoroughly examined, and the conclusions should be described in well-written documents.

8.4.6 Software Quality Assurance

Because software is an important element of a medical device, a software quality assurance program is essential when a design incorporates software developed in-house. To ensure overall functional reliability of the device software, the program should include a protocol for formal review and validation. The main objectives of the software quality assurance program should be reliability, correctness, maintainability, and testability.

The program should also assure (if applicable) that the subcontractor has a satisfactory software quality assurance program so that the software reliability, correctness, maintainability, and testability are effectively looked after.

8.4.7 Labelling

This includes display labels, panels, charts, manuals, inserts, and recommended test and calibration protocols. The design review process should also review labelling to assure it complies with applicable laws and regulations and contains easy-to-understand directions. The verification of the accuracy of instructions contained in the labelling should be a part of the qualification testing of the device under consideration.

Maintenance manuals (if applicable) must be written clearly so that the device could be maintained in an effective and safe condition.

8.4.8 Design Transfer

After translating the design into a physical entity, the design's technical adequacy, reliability, and safety should be verified through intensive testing under simulated or real-life use environments. After verifying technical adequacy through appropriate testing, the design is usually approved. Note that when moving from laboratory to scaled-up production, standards, or methods and procedures may not be effectively transferred. It is possible that additional manufacturing processes are needed. Thus, this scenario requires careful consideration.

8.4.9 Certification

After the successful passing of preproduction qualification testing by the initial production units, it is appropriate to conduct a formal technical review so that the adequacy of the design, production, and quality assurance procedures is assured. In addition, the review should determine the following factors:

- Adequacy of the specification change control program
- Adequacy of the total quality assurance plan
- Suitability of test methods employed to evaluate compliance with respect to approved specifications
- Adequacy of specifications
- Resolution of any discrepancy between the final approved device specifications and the actual end device
- Resolution of any discrepancy between the standards and procedures employed to produce the design during research and development and those recommended for the production phase

8.4.10 Test Instrumentation

This involves effectively calibrating and maintaining all equipment used in the qualification of the design. More specifically, such equipment should be kept under a formal calibration and maintenance program.

8.4.11 Personnel

This calls for the performance of design activities, including design review, analysis, and testing by properly trained professionals.

8.4.12 Quality Monitoring After the Design Phase

The effort to produce safe, reliable, and effective medical devices does not end when the design phase is completed; it continues during the manufacturing and field use phases. Thus, medical device manufacturers should have an effective program for purposes such as the following:

- To identify failure patterns.
- To analyze quality problems.
- To take appropriate corrective measures to prevent recurrence of identified problems.
- To have timely internal reporting of problems found either in-house or in the field environment.

Device manufacturers should also make a special effort to assure that failure data collected from service and complaint records relating to design problems are reviewed by the design professionals.

8.5 TOTAL QUALITY MANAGEMENT

Total quality management (TQM) is an advancement to the traditional approach of doing business. It may simply be stated as the art of managing the whole to achieve excellence.[9] More specifically, it is a quality emphasis that encompasses the whole organization, including suppliers and customers. The term "total quality management" was coined by Nancy Warren, a behavioral scientist, in 1985.[10]

TQM can equally be applied in the manufacture of medical devices, and it will certainly help to produce safe, reliable, and effective medical items. Basically, there are seven important elements of TQM, as shown in Figure 8.1.[11–12] These are team effort, management commitment and leadership role, customer service, cost of quality, training, supplier participation, and statistical methods and techniques.

For the effectiveness of the TQM process, some of the objectives that must be satisfied are as follows:

- Total understanding of requirements (i.e., of internal and external customer) by everyone in the organization
- Meeting control guidelines, as specified by the customer requirements, by all significant systems and processes
- Establishing rewards and incentives for workers when customer satisfaction and process control results are effectively achieved
- Using a mechanism that always improves processes, which in turn better satisfy current and potential requirements of customers

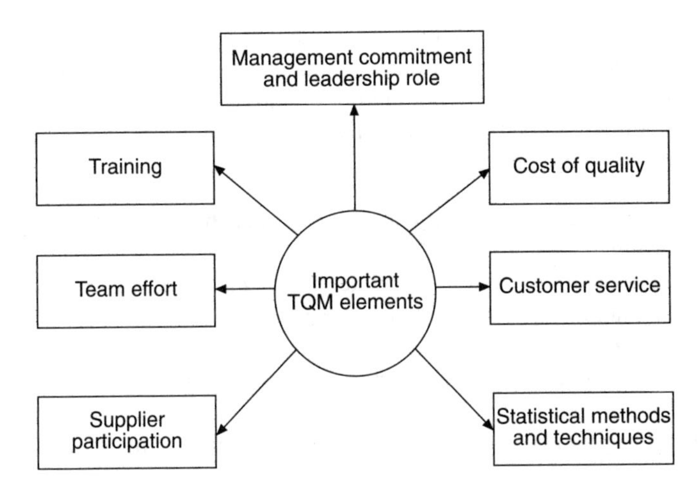

FIGURE 8.1 Important elements of TQM.

8.5.1 Obstacles to TQM Implementation

The implementation of TQM is a challenging task. Clear answers to questions on areas such as those listed below will help reduce many associated obstacles:[13]

- Understanding of the TQM purpose by the management
- Setting the TQM vision
- Senior management's support in the introduction of TQM
- Mechanism to quantify customer requirements
- Convincing individuals of the need to change
- Convincing individuals that TQM is different
- Time available to implement TQM programs

8.6 TOOLS FOR ASSURING MEDICAL DEVICE QUALITY

Over the years, many tools have been developed for use in quality work.[14] Such tools can be applied equally to assure quality of medical devices. According to K. Ishikawa, 95% of quality-related problems can be resolved by using seven basic tools: Pareto diagram, flowcharts, check sheets, cause-and-effect diagrams, control charts, scatter diagrams, and histograms.[15] All of these methods, in addition to quality function deployment, are described below.[9,12,14-16]

8.6.1 Pareto Diagram

This is named after Italian economist Vilfredo Pareto (1848–1923) who extensively studied the distribution of wealth in Europe and concluded that a large percentage of wealth is owned by a small percentage of the population. However, Joseph Juran, one of the quality gurus, recognized its application in quality work and stated that 80% of quality problems are the result of only 20% of the causes.

A Pareto diagram (i.e., a type of frequency chart) arranges data in a hierarchical order, thus helping to identify the most significant problems to be corrected first. Although the Pareto approach can summarize all types of data, it is employed basically to identify and determine nonconformities. The main steps involved in the construction of a Pareto diagram are[9]

- Determine the approach to classify data, i.e., by cause, problem, nonconformity.
- Decide what to use to rank characteristics, i.e., frequency or dollars.
- Obtain data for appropriate time intervals.
- Summarize the data and rank classifications from largest to smallest.
- Construct the diagram and determine the significant few.

This diagram could be extremely useful to improve the quality of medical device designs.

8.6.2 Cause-and-Effect Diagram

This diagram was developed by Kaoru Ishikawa in 1943, thus sometimes it is also called the Ishikawa diagram. Another name used for the diagram is "Fishbone diagram" because of its resemblance to the bones of a fish. The diagram shows a desirable or undesirable outcome as an effect, and related causes as leading to or potentially leading to that effect. Thus, the diagram can be used to investigate either (i) a "bad" effect and taking appropriate measures to rectify the causes, or (ii) a "good" effect and learning about the causes responsible for it. For example, in the cause-and-effect diagram for the TQM effort, the effect could be customer satisfaction and the major causes could be methods, manpower, materials, and machines. The analysis of such causes can serve as an effective tool to identify possible quality-related problems and inspection points.

Figure 8.2 shows the main steps involved in developing a cause-and-effect diagram. Visually, the right side of the diagram, or the "fish head," denotes effect, and its left side shows all possible causes linked to the central "fish" spine.

There are many benefits of the cause and effect diagram, including the fact that it is useful to generate ideas and highlight the root cause; it presents an orderly arrangement of theories; and it is an effective tool to guide further inquiry.

Its main disadvantage is that users can overlook critical, complex interactions between causes. For example, if a problem is due to a combination of factors, the cause-and-effect diagram is a difficult tool to depict and solve it.

8.6.3 Scatter Diagram

This is the simplest way to determine how two variables are related or if a cause-and-effect relationship exists between the two variables. However, the scatter diagram cannot prove that one variable causes the change in the other, but only the existence of their relationship and its strength. In this diagram, the horizontal axis denotes the measurement values of one variable and the vertical axis denotes the measurements of the other variable.

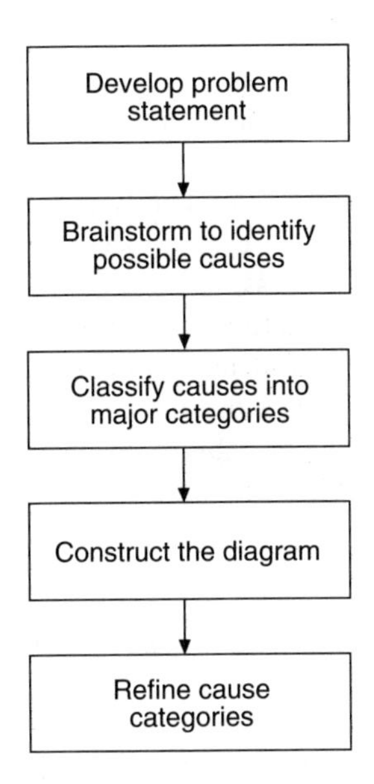

FIGURE 8.2 Major steps for developing a cause-and-effect diagram.

If sometimes it is desirable to fit a straight line to the plotted data points to obtain a prediction equation, a line can be drawn on the scatter diagram either visually or mathematically using the least squares approach. Whenever the line is extended beyond the plotted data points, a dashed line is used to indicate that there are no data for the concerned area.

8.6.4 Check Sheets

These are used to collect data effectively so that such data can easily and efficiently be used and analyzed. More specifically, check sheets are useful to organize data by classification and indicate the number of times each specific value occurs. This information becomes increasingly important and useful as more data are collected; however, there must not be fewer than 50 observations available, to be charted, for this approach to be effective. Some important points associated with the check sheets are

- Check sheets are useful to operators to spot problems when the frequency of a specific defect and how frequently it occurs in a particular location are shown.
- The check sheet form is individualized for each situation, and usually the project team designs it.
- Creativity is important during the design of a check sheet.
- Always aim to make the check sheet user-friendly and include data on time and location.

8.6.5 Histogram

A histogram is used when good clarity is desired. It plots data in a frequency distribution table. Its main distinction from a check sheet is that its data are classified into rows for the purpose of losing the identity of individual values. It may be said that the histogram is the first "statistical" process control technique because it can describes the variation in the process.

A histogram can give a satisfactory amount of information concerning a quality problem, thus providing a basis for making decisions without additional analysis. The shape of the histogram shows the nature of the distribution of the data, in addition to central tendency and variability. Furthermore, specification limits may be used to show process capability.

8.6.6 Flowcharts

Flowcharts are used to describe processes in as much detail as feasible by graphically showing the steps in proper order. A good flowchart usually displays all process steps under consideration or analysis by the quality improvement team, highlights crucial process points for control, suggests areas for improvement, and serves as a useful tool to explain and solve a problem.

A flowchart could be simple or complex, composed of many boxes, symbols, etc. More specifically, the complex version indicates the process steps in the appropriate sequence and related step conditions and the associated constraints by making the use of elements such as arrows, if/then statements, or yes/no choices.

8.6.7 Quality Function Deployment (QFD)

Quality Function Deployment was developed by Professor Mizuno of the Tokyo Institute of Technology, and its first application was at Mitsubishi Heavy Industries Ltd. in 1972.[9,14,17] In 1984, QFD was applied for the first time in the U.S. by the Xerox Corp. QFD may simply be described as a planning tool used to satisfy customer expectations. Furthermore, it is a systematic mechanism to item design and manufacture and provides in-depth evaluation of a product or item. QFD emphasizes customer requirements or expectations and is frequently called the voice of the customer. It uses a set of matrices to relate customer expectations to counterpart characteristics expressed as technical specifications and process control requirements. In a nutshell, the customer- or consumer-need planning matrix forms a critical element of the QFD approach. Often, because of its resemblance to a house, QFD is called "The House of Quality." Some of the main steps associated with QFD are as follows:

- Identify consumer expectations.
- Highlight the product characteristics that will meet the consumer's requirements.
- Associate consumer requirements and counterpart characteristics.
- Evaluate competing products.

- Evaluate counterpart characteristics of competing products and develop objectives.
- Select counterpart characteristics to be used in the remaining process.

Some of the benefits of QFD include improvement in engineering knowledge, productivity, and quality and reduction in cost, engineering changes, and product development time. In contrast, its important drawback is that the exact needs must be identified in complete detail.

8.6.8 Control Charts

In quality control work, various types of control charts are used to monitor the state of processes. A control chart simply shows statistically calculated upper and lower limits on either side of a process mean value. More specifically, the control chart displays if the collected data are within upper and lower limits calculated previously by using raw data obtained from earlier trials.

Note that the basis for the construction of a control chart is statistical principles and distributions, in particular, the normal distribution. When employed in conjunction with a manufacturing process, the control chart is extremely useful to indicate trends and signal when the process goes out of control. The results of the process are monitored over a period of time. When they are not within the specified control limits, an investigation is performed to determine the cause and subsequently corrective measures are taken.

All in all, a control chart is useful to determine variability and its reduction as much as economically possible.

8.7 QUALITY INDICES

Over the years, there have been numerous indices developed to aid product manufacturers. This section presents some indices that could be useful to medical device or equipment manufacturers.

8.7.1 Vendor Rating Program Index

This index concerns evaluating supplier quality cost performance by using the quality cost and is defined by[18]

$$QCI = \frac{VQC + PC}{PC}, \tag{8.1}$$

where

QCI is the value of the quality cost performance index.
PC is the purchased cost.
VQC is the vendor quality cost.

The value of this index equals only unity when the vendor is perfect, i.e., there is no vendor quality cost. For example, there is no complaint to investigate, no receiving inspection, no defective rejection, etc. When the value of QCI is 1.1 or greater, it indicates that there is an urgent need for the corrective action. Interpretations for other values of the index are as follows:[18]

- $1.000 < VQC < 1.009$ Excellent performance
- $1.010 < VQC < 1.03$ Good performance

Example 8.1

Assume that we have the following data:

- PC = $80,000
- VQC = $3000

Calculate the value of the quality cost performance index, and comment on the result.

By substituting the given data into Equation (8.1), we get

$$QCI = (3000 + 80,000)/(80,000)$$
$$= 1.0375$$

It means the quality cost performance of the vendor can be rated as fair.

8.7.2 Quality Cost Index

This index involves measuring a manufacturer's quality cost performance and is defined by[19]

$$\alpha = \frac{QC}{VO}(100) + 100, \tag{8.2}$$

where

α is the value of the quality cost index.
QC is the total quality cost.
VO is the total value of output.

The value of this index may be estimated in six steps:

1. Establish time base, i.e., quarter, month, etc.
2. Determine the value of the total output, i.e., the total value of all finished products of an acceptable level of quality.
3. Calculate the value of the scrap produced or generated.

4. Estimate the total value of labor costs, for example, cost of quality control inspection.
5. Add the results of the previous two steps (steps 3 and 4).
6. Compute the value of the quality cost index, α, by using Equation (8.2) and the resulting values of steps 2 and 5.

Interpretations of some α values are as follows:

- $\alpha = 100$: There is no defective output, thus no money is spent to perform quality checks.
- $\alpha = 105$: This value is achievable in real life.
- $100 < \alpha < 130$: This is the common range when quality costs are ignored by manufacturers.

8.7.3 Quality Inspector Accuracy Index

Because quality inspectors can accept bad items and reject good ones, the check inspectors may be used to reexamine both the accepted and rejected items. Thus, this index concerns measuring the accuracy of regular inspectors. The index is defined by[20]

$$\theta = \frac{\lambda - \mu}{\lambda - \mu + b}(100), \tag{8.3}$$

where

- θ is the percentage of defects accurately identified by the regular inspector.
- b is the number of defects missed by the regular inspector as discovered by the check inspector.
- λ is the total number of defects found by the regular inspector.
- μ is the total number of items or units without defects rejected by the regular inspector as discovered by the check inspector.

Example 8.2

Assume a lot of medical devices was inspected by a regular inspector who discovered 40 defects. Subsequently, the same lot was reexamined by the check inspector, and the values of μ and b were 5 and 8, respectively. Determine the percentage of defects accurately found by the regular inspector.

By inserting the specified values into Equation (8.3), we get

$$\theta = \frac{(40-5)}{(40-5)+8}(100) = 81.395\%.$$

It means that the percentage of defects accurately found by the regular inspector is 81.395%.

8.7.4 Quality Inspector Inaccuracy Index

This index concerns calculating the percentage of good items or devices rejected by the regular inspector and is defined by[21]

$$\beta = \frac{\mu(100)}{\theta - (\lambda - \mu + b)}, \tag{8.4}$$

where

θ is the total number of items or devices inspected.

8.8 PROBLEMS

1. Discuss historical developments in quality control in general and medical device quality assurance in particular.
2. Define the following terms:
   - Quality assurance
   - Inspection
   - Quality
   - Quality measure
3. Write an essay on ISO 9000.
4. What are the elements of the FDA's "Preproduction Quality Assurance Planning: Recommendations for Medical Device Manufacturers" program?
5. Discuss in detail at least five elements listed in question 4.
6. What are the important elements of TQM?
7. Discuss the following tools that can be used to assure quality of medical devices:
   - Quality function deployment
   - Pareto diagram
   - Cause-and-effect diagram
8. What are the major steps involved in developing a cause-and-effect diagram?
9. Assume that we have the following data:
   - PC = $70,000
   - VQC = $2000

   Compute the value of the quality cost performance index using Equation (8.1), and comment on the end result.
10. A lot containing medical components was inspected by a regular inspector who found 50 defects. Later, the same lot was reexamined by the check inspector, and the values of μ and b were 7 and 9, respectively. Calculate the percentage of defects correctly discovered by the regular inspector using Equation (8.3).

REFERENCES

1. Dague, D.C., Quality: Historical Perspective, in *Quality Control in Manufacturing*, Published by the Society of Automotive Engineers Inc., Warrendale, PA, 1981.
2. Fagan, M.D., Ed., *A History of Engineering and Science in the Bell System, the Early Years (1875–1925)*, Published by the Bell Telephone Laboratories Inc., New York, 1974.
3. Banks, J., *Principles of Quality Control*, John Wiley & Sons, Inc., New York, 1989.
4. Omdahl, T.P., Ed., *Reliability, Availability, and Maintainability (RAM) Dictionary*, ASQC Quality Press, Milwaukee, WI, 1988.
5. Fries, R.C., *Medical Device Quality Assurance and Regulatory Compliance*, Marcel Dekker Inc., New York, 1998.
6. Dhillon, B.S., *Quality Control, Reliability, and Engineering Design*, Marcel Dekker Inc., New York, 1985.
7. Montanez, J., *Medical Device Quality Assurance Manual*, Interpharm Press Inc., Buffalo Grove, IL, 1996.
8. Hooten, W.F., A Brief History of FDA Good Manufacturing Practices, *Med. Dev. Diag. Ind. Mag.*, Vol. 18, No. 5, 1996, p. 96.
9. Besterfield, D.H., Besterfield-Michna, C., Besterfield, G.H., and Besterfield-Sacre, M., *Total Quality Management*, Prentice-Hall Inc., Englewood Cliffs, NJ, 1995.
10. Walton, M., *Deming Management at Work*, Putnam, New York, 1990.
11. Burati, J.L., Matthews, M.F., and Kalidindi, S.N., Quality Management Organizations and Techniques, *J. Const. Eng. Manag.*, Vol. 118, March 1992, pp. 112–128.
12. Dhillon, B.S., *Advanced Design Concepts for Engineers*, Technomic Publishing Company, Lancaster, PA, 1998.
13. Klein, R.A., Achieve Total Quality Management, *Chemical Engineering Progress*, November 1991, pp. 83–86.
14. Mears, P., *Quality Improvement Tools and Techniques*, McGraw-Hill Inc., New York, 1995.
15. Sahni, A., Seven Basic Tools That Can Improve Quality, *Med. Dev. Diag. Ind. Mag.*, April 1998, pp. 89–98.
16. Bracco, D., How to Implement a Statistical Process Control Program, *Med. Dev. Diag. Ind. Mag.*, March 1998, pp. 129–139.
17. Yoji, K., Ed., *Quality Function Deployment*, Productivity Press, Cambridge, MA, 1990.
18. Guide for Managing Vendor Quality Costs, American Society for Quality Control, Milwaukee, WI, 1980.
19. Lester, R.H., Enrick, N.L., and Mottley, H.E., *Quality Control for Profit*, Industrial Press Inc., New York, 1977.
20. Juran, J.M. Gryna, F.M., and Bingham, R.S., *Quality Control Handbook*, McGraw-Hill, New York, 1974.

9 Medical Device Reliability Testing

CONTENTS

9.1 INTRODUCTION

Testing is an important factor in any engineering product development program, and it may be described as subjecting the product to conditions that highlight a product's modes of failure, behavior characteristics, and weaknesses. There could be many purposes for testing, including comparing vendors, determining the design feasibility, developing reliability parameters, comparing configurations, and determining the response to environmental stresses.

Testing is conducted in two modes: standard or accelerated. In the case of standard mode, the tests are performed at ambient temperature and at typical operating parameters, and the actual time of operation is considered as the test time. In the accelerated mode, parameters such as temperature, frequency of cycling, or voltage are varied above their normal levels to reduce test time, or it could simply be a test such as sudden death testing.

Reliability testing is an important element of testing, and its main objective is to obtain information regarding failures, particularly, the product/device's tendency to fail and the consequences of the failure. A good reliability test program requires minimal testing and provides the maximum amount of information regarding failures. Usually, reliability testing is performed according to two U.S. military documents.[1,2] A comprehensive list of publications on reliability testing is given in reference 3.

In medical devices, reliability testing is a crucial factor. This chapter discusses aspects of reliability testing, along with related areas useful for medical devices.

9.2 TEST PLAN INFORMATION AND TEST TYPES

A plan is an essential element of testing, and without it, testing is not testing at all but simply an experiment. Each test performed must be described in detail in the test plan, and the plan should include information such as the description of the device or item involved, the type of test and its purpose, any special requirements and parameters to be recorded, the test length and the number of devices or units on test, description of the approach for running the test, and the definition of each failure.[4,5]

Various types of tests are performed, as shown in Figure 9.1, to obtain information concerning an item's behavior characteristics, failure modes, weaknesses, etc. Each test classification shown in Figure 9.1 is discussed below.[4,5]

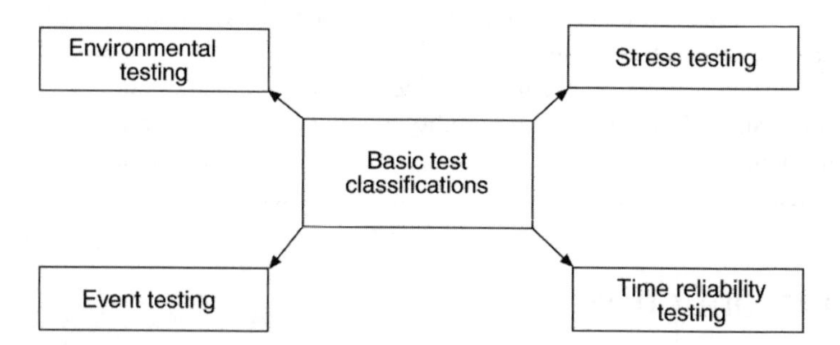

FIGURE 9.1 Basic test classifications.

9.2.1 Event Testing

This is an important type of testing. It is analogous to time to failure testing and involves repeated testing of a device or equipment through its cycle of operation until the occurrence of failure. The number of cycles to failure is the parameter developed from the event testing.

9.2.2 Time Reliability Testing

The primary purpose of this type of testing is to estimate the values of parameters such as mean time between failures and failure rate. Also, to establish priorities of criticality for reliability improvement, time or long-term reliability testing can be performed to determine factors such as the following:

- Which element/part fails?
- When does the element/part fail?
- What is the mode of failure at that specific point in time?
- What is the mechanism of failure?
- What is the approximate remaining life of the equipment/device that is required for field use?

9.2.3 Stress Testing

This type of testing is important in reliability assessment of engineering items or devices, but its application requires careful consideration because too much stress may lead to inclusive results. The over-stress should be performed in steps rather than reaching to the maximum value immediately because in the event of device or item failure, the step approach allows one to determine where in the progression the failure occurred.

9.2.4 Environmental Testing

This is performed to assure the item's ability to withstand the environmental stresses during shipping and use. Before starting any environment test, the item under consideration is tested mechanically and electrically to assure that it is operating according to the design specification. After each environment test, the item is again tested electrically and mechanically to ascertain if the environment test has caused any change in its specified operation. Environmental testing may be classified into categories such as the following:

- **Thermal shock testing**. This concerns assuring that the item or device under consideration can withstand the stresses associated with alternate exposure to hot and cold temperatures.
- **Mechanical shock testing**. This assures that the item or device under consideration can withstand stresses of shipping, handling, and day-to-day use. Furthermore, the testing may be performed when the item is either packaged or unpackaged.
- **Operating temperature testing**. This assures that the item or device under consideration will function according to the specification at the extremes of the normal operating temperature range. Furthermore, this type of testing is quite useful to perform analysis of the internal temperatures of the item under consideration to assure that none exceed the temperature limits of any parts.

- **Humidity testing**. This assures that the item or device under consideration can withstand the stresses associated with exposure to a humid environment.
- **Storage temperature testing**. This assures that the item or device under consideration can withstand the stresses associated with shipping and storage.
- **Impact testing**. This assures that the item or device under consideration can withstand the collision stresses associated with shipping and day-to-day use of mobile items or devices.
- **Electromagnetic compatibility testing**. This determines the maximum amount of electromagnetic emissions the item or device under consideration is permitted to generate, as well as ascertains the lowest level of electromagnetic interference to which the item in question must not be susceptible. It is important to perform electromagnetic compatibility testing on items such as microprocessor-based devices, circuits with clock or crystal oscillators, and monitors that are used in close proximity to other devices.
- **Electrostatic discharge testing**. This assures that the item or device under consideration can withstand short duration voltage transients due to factors such as static electricity, inductive or capacitive effects, and load switching.
- **Mechanical vibration testing**. This assures that the item or device under consideration can withstand the vibration-related stress of shipping, handling, and day-to-day use, in particular where the item is mobile.

9.3 CLASSIFICATIONS OF RELIABILITY TESTS

To obtain various types of information, many types of reliability tests are performed. They may be grouped under three classifications:[6]

- Reliability development and demonstration tests
- Qualification and acceptance tests
- Operational tests

There are three main objectives of reliability development and demonstration tests: (i) to determine if there is any need to improve design to satisfy the reliability specification, (ii) to verify improvements occurring in design reliability, and (iii) to identify any required changes in design. Note that the objectives of reliability development and demonstration tests are closely related to the development of the prototype model, and such tests are usually handled by professionals in charge of item or device design and development.

Qualification and acceptance tests have two-fold objectives: (i) to assess if a specific design under consideration should be considered as qualified for its stated goal and (ii) to determine if the item should be accepted or rejected either individually or on lot basis. Note that the objectives of qualification and acceptance tests differ from those of other reliability tests, especially with regard to the accept-or-reject approach.

Operational tests are conducted to provide the final verification of design and permit the experimental observation of the relationship between field experience and the individual segments of a reliability effort expended. More specifically, these tests are useful to provide feedback from field experience to theory. Operational tests have three-fold objectives: (i) verifiying the reliability analyses conducted during the project, (ii) providing data indicating appropriate modifications of operational policies and procedures as they influence item reliability and maintainability, and (iii) providing useful information to be utilized in subsequent activities.

9.4 ACCELERATED TESTING

This may be described as reducing the length of the test time by varying parameters such as temperature, frequency of cycling, or voltage above their regular levels or simply, for example, performing sudden death testing. The availability of the test time dictates whether the test is conducted in an accelerated or a standard mode. There are several ways to accelerate testing, including the three shown in Figure 9.2.[4,7-13]

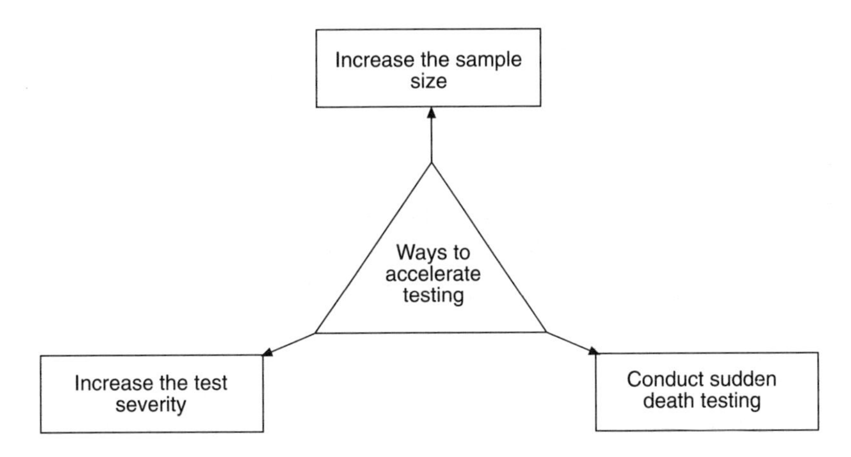

FIGURE 9.2 Ways to accelerate testing.

Increasing the sample size is one way to accelerate reliability test when the item life distribution does not exhibit a wearout characteristic during the predicted lifespan of the item. In this case, as the test time is inversely proportional to the size of the sample, the increase in the sample size decreases the test time.

In sudden death testing, the entire test sample is divided into a number of groups containing an equal number of units or items, and all the units in each group start their operation simultaneously. The entire group of units is considered failed whenever one of its units fails and the testing of the remaining unfailed units is terminated. Furthermore, at the failure of the first unit in the last group, the entire testing process is stopped. Another approach to acceleration testing is to increase the test severity.

This is a logical approach to reduce test time when it is impossible to use large sample sizes. The approach involves increasing the severity of tests by increasing the stress acting on the test item. The stress may be grouped into two areas: operational and application. The operational area includes temperature and humidity, and the application area includes current, voltage, self-generated heat, etc.

The following three additional factors are also associated with this type of accelerated testing:

- The common approach to accelerated testing is to increase the temperature severity.
- Ensure that the higher stresses do not introduce unrealistic failure modes.
- It is quite possible that the separate stresses may interact, and their combined weakening effect is larger than expected from a single additive process.

For this type of accelerated testing, the acceleration factor may be calculated by using the following equation:[4,12-13]

$$AF = \exp\left[-(E/k)\left(\frac{1}{T_a} - \frac{1}{T_u}\right)\right], \tag{9.1}$$

where

AF	is the acceleration factor.
k	is Boltzman's constant, and its value is taken as 0.00008623 eV/degree Kelvin.
E	is the energy of activation and its value is taken as 0.5 eV.
T_a	is the accelerate temperature in degrees Kelvin.
T_u	is the use temperature in degrees Kelvin.

Example 9.1

Assume that 15 identical medical devices were tested to failure at the 140°C, and their mean time between failures (MTBF) was 3000 hours. Calculate the MTBF of the medical devices at the normal operating temperature of 80°C.

Thus, we have

$$T_u = 273 + 80 = 353 \text{ degrees Kelvin}$$

$$T_a = 273 + 140 = 413 \text{ degrees Kelvin}$$

By substituting the given and other data into Equation (9.1), we get

$$AF = \exp\left[-\left(\frac{0.5}{0.00008623}\right)\left\{\frac{1}{413} - \frac{1}{353}\right\}\right]$$

$$= 10.87$$

Thus, the MTBF of the medical devices at 80°C is

$$\begin{aligned} MTBF_n &= (10.87)(3000) \\ &= 32{,}610 \text{ hours}, \end{aligned}$$

where

$MTBF_n$ is the mean time between failures at 80°C.

9.4.1 Some Relationships Between Accelerated and Normal Operating Conditions

Under a true linear acceleration, we can write down the following relationships between the accelerated and normal operating conditions:[12,13]

Time to Failure

The following equation describes the relationship between the time to failure at normal operating condition and stress condition:

$$T_N = (AF)\,T_S, \tag{9.2}$$

where

- T_S is the time to failure under stress condition.
- T_N is the time to failure under normal condition.

Probability Density Function

The following equation gives the relationship between the probability density function at normal operating condition and stress condition:

$$f_N(t) = \frac{1}{AF} f_S\left(\frac{t}{AF}\right), \tag{9.3}$$

where

- t is time.
- $f_N(t)$ is the failure probability density function at the normal operating condition.
- $f_S\left(\frac{t}{AF}\right)$ is the failure probability density function at the stressful operating condition.

Cumulative Distribution Function

The following equation provides the relationship between the cumulative distribution function at normal operating condition and stress condition:

$$F_N(t) = F_S\left(\frac{t}{AF}\right), \quad (9.4)$$

where

$F_N(t)$ is the cumulative distribution function at the normal operating condition.

$F_S\left(\frac{t}{AF}\right)$ is the cumulative distribution function at the stressful operating condition.

Hazard Rate

The following equation describes the relationship between the hazard rate at normal operating condition and stress condition:

$$\lambda_N(t) = \frac{f_N(t)}{1 - F_N(t)} = \frac{1}{AF}\lambda_S\left(\frac{t}{AF}\right), \quad (9.5)$$

where

$\lambda_N(t)$ is the hazard rate at the normal operating condition.

$\lambda_S\left(\frac{t}{AF}\right)$ is the hazard rate at the stressful operating condition.

Example 9.2

Assume that time to failure of a medical device under accelerated stress is exponentially distributed as described by the following probability density function:

$$f_S(t) = \lambda_S e^{-\lambda_S t}, \quad (9.6)$$

where

λ_S is the distribution parameter or the constant failure rate at the stressful level.

Obtain expressions for the cumulative distribution function and the hazard rate at the normal operating condition.

Thus, substituting Equation (9.6) into the Equation (9.4), we get

$$F_N(t) = F_S\left(\frac{t}{AF}\right) = 1 - e^{-\left(\frac{\lambda_S t}{AF}\right)} \quad (9.7)$$

By using Equations (9.3)–(9.4) and (9.6)–(9.7) in Equation (9.5), we get

$$\lambda_N(t) = \frac{\lambda_S}{AF} \quad (9.8)$$

Thus, expressions for the cumulative distribution function and the hazard rate at the normal operating condition are given by Equations (9.7) and (9.8), respectively.

9.5 SUCCESS TESTING

This is an important type of testing that can be used in the testing of medical devices, and it is practiced in receiving inspection and in engineering test laboratories where no-failure test is specified. Normally, the primary objective of this type of testing is to ensure that a specified reliability level is achieved at a stated level of confidence. Thus, for zero failures, the lower 100 $(1-\alpha)$ percent confidence limit on the required level of reliability can be expressed as follows:[16,17]

$$R_L = \alpha^{\frac{1}{n}}, \tag{9.9}$$

where

α is the consumer's risk (i.e., the level of significance)
n is the total number of items placed on test.

Thus, with 100 $(1-\alpha)$ percent confidence, we may write

$$R_L \leq R_A, \tag{9.10}$$

where

R_A is the actual or true reliability.

By taking the natural logarithms of both sides of Equation (9.9), we get

$$\ln R_L = \frac{1}{n}\ln \alpha \tag{9.11}$$

Rearranging Equation (9.11) yields

$$n = \frac{\ln \alpha}{\ln R_L} \tag{9.12}$$

The desired level of confidence, DLC, may be expressed as follows:

$$DLC = 1 - \alpha \tag{9.13}$$

Rearranging Equation (9.13), we get

$$\alpha = 1 - DLC \tag{9.14}$$

Using Equations (9.10) and (9.14) in Equation (9.12) yields

$$n = \frac{\ln(1 - DLC)}{\ln R_A} \tag{9.15}$$

The above equation can be used to determine the total number of units or items to be placed on test for given reliability and level of confidence.

Example 9.3

A manufacturer of pacemakers is required to demonstrate 99% reliability of a pacemaker at 80% confidence level. Determine the total number of pacemakers to be tested when zero failures are permitted.

By substituting the specified data into Equation (9.15), we get

$$n = \frac{\ln(1 - 0.80)}{\ln 0.99}$$

$$\simeq 160 \text{ pacemakers}$$

It means a total of 160 pacemakers must be tested.

9.6 MEAN TIME BETWEEN FAILURES (MTBF) CALCULATION METHODS IN RELIABILITY TESTING

MTBF is an important parameter that provides useful information regarding the reliability of an item. There are basically five ways to estimate MTBF in reliability testing as shown in Figure 9.3.

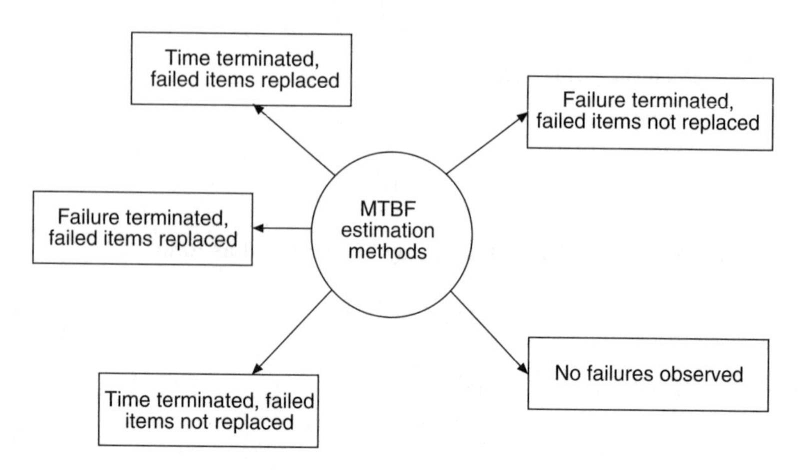

FIGURE 9.3 Methods to estimate MTBF in reliability testing.

Each method is described below.[4,5]

9.6.1 Failure Terminated, Failed Items Not Replaced

In this case, a specific number of items is placed on test and the testing is terminated at a preassigned number of failures. Furthermore, the failed items are not replaced. The MTBF is expressed by

$$\text{MTBF} = \left\{ \sum_{i=1}^{n} T_i + (M - n)T_d \right\} \Big/ n, \tag{9.16}$$

where

- T_i is the time to failure.
- M is the number of items placed on the test.
- T_d is the test duration time.
- n is the total number of failures.

Example 9.4

Ten identical units of a medical equipment were put on test at time T = 0, and five of the units failed after 80, 100, 140, 160, and 200 hours. After the failure of the fifth unit (at 200 hours), the test was terminated and none of the failed units was replaced. Calculate the MTBF of the units tested.

Substituting the given data into Equation (9.16), we get

$$\begin{aligned} \text{MTBF} &= \left[(80 + 100 + 140 + 160 + 200) + (10 - 5)200\right]/5 \\ &= 336 \text{ hours} \end{aligned}$$

Thus, the MTBF of the units tested is 336 hours.

9.6.2 Time Terminated, Failed Items Not Replaced

In this case, the testing is terminated at a preassigned time, and the failed units or items are not replaced. Thus, the MTBF is given by

$$\text{MTBF} = \left\{ \sum_{i=1}^{n} T_i + (M - n)T_d \right\} \Big/ n \tag{9.17}$$

Example 9.5

A manufacturer placed eight identical ventilators on test at zero time and tested them for 1000 hours. Four of the ventilators failed after 250, 400, 650, and 800 hours, and none of the failed units was replaced. Determine the MTBF of the ventilators placed on test.

Inserting the specified data into Equation (9.17) yields

$$\text{MTBF} = \{(250 + 400 + 650 + 800) + (8 - 4)(1000)\}/4$$
$$= 1525 \text{ hours}$$

Thus, ventilator MTBF is 1525 hours.

9.6.3 Failure Terminated, Failed Items Replaced

In this case, the testing is stopped at a preassigned number of failures, and each failed item is replaced. Thus, the MTBF is expressed by

$$\text{MTBF} = \text{MT}_d/\text{n} \tag{9.18}$$

Example 9.6

Assume that in Example 9.4 each failed unit was replaced. Calculate the MTBF of the units placed on test.

By substituting the given data into Equation (9.18), we get

$$\text{MTBF} = (10)(200)/4$$
$$= 400 \text{ hours}$$

Thus, the MTBF of the units tested is 400 hours.

9.6.4 Time Terminated, Failed Items Replaced

In this case, the testing is terminated at a preassigned time, and all failed items are replaced after their individual failure. Thus, the MTBF is given by

$$\text{MTBF} = \text{MT}_d/\text{n} \tag{9.19}$$

Example 9.7

Assume that in Example 9.5 each failed ventilator is replaced. Calculate the MTBF of the ventilators tested.

Inserting the specified data into Equation (9.19) yields

$$\text{MTBF} = (8)(1000)/4$$
$$= 2000 \text{ hours}$$

Thus, the MTBF of the ventilators is 2000 hours.

9.6.5 No Failures Observed

In this case, MTBF cannot be calculated because no failures are observed. However, a lower one-sided confidence limit can be computed to state the minimum limit of the MTBF for a specified confidence level. Thus, the lower one-sided confidence limit is given by

$$\mathrm{LOCL} = \left(2\mathrm{T_t}\right)/\chi^2_{\alpha,\mathrm{m}} \tag{9.20}$$

where

LOCL is the lower one-sided confidence limit.

$\chi^2_{\alpha,m}$ is the chi square distribution value, where α is the level of risk and *m* the degrees of freedom, and in this case its value is equal to 2.

T_t is the total test time.

Example 9.8

Assume that 12 identical pacemakers were placed on test for 1500 hours and none of them failed. Determine the minimum limit of the pacemaker MTBF at the 95% level of confidence.

Thus, the total duration of the test is

$$= (1500)(12)$$

$$= 18,000 \text{ hours}$$

The level of risk, α, is given by

$$\alpha = 1 - 0.95$$

$$= 0.05$$

By substituting the given data and the calculated values into Equation (9.20), we get

$$\mathrm{LOCL} = 2(18,000)/\chi^2_{0.05,2} \tag{9.21}$$

Using Table 9.1 and Equation (9.21) yields

$$\mathrm{LOCL} = 2(18,000)/(5.99)$$

$$= 6010 \text{ hours}$$

Thus, the minimum value of the pacemaker MTBF at the 95% confidence level is 6010 hours.

TABLE 9.1
Chi-Square Distribution Values[18]

Degrees of Freedom (m)	Probability									
	0.99	**0.95**	**0.90**	**0.80**	**0.70**	**0.50**	**0.2**	**0.10**	**0.05**	**0.01**
2	0.02	0.10	0.21	0.44	0.71	1.38	3.21	4.60	5.99	9.21
3	0.11	0.35	0.58	1.00	1.42	2.36	4.64	6.25	7.81	11.34
4	0.29	0.71	1.06	1.64	2.19	3.35	5.98	7.77	9.48	13.27
5	0.55	1.14	1.61	2.34	3.00	4.35	7.28	9.23	11.07	15.08
6	0.87	1.63	2.20	3.07	3.82	5.34	8.55	10.64	12.59	16.81
7	1.23	2.16	2.83	3.82	4.67	6.34	9.80	12.01	14.06	18.47
8	1.64	2.73	3.49	4.59	5.52	7.34	11.03	13.36	15.50	20.09
9	2.08	3.32	4.16	5.38	6.39	8.34	12.24	14.68	16.91	21.66
10	2.55	3.94	4.86	6.17	7.26	9.34	13.44	15.98	18.30	23.20
12	3.57	5.22	6.30	7.80	9.03	11.34	15.81	19.54	21.02	26.21
14	4.66	6.57	7.79	9.46	10.82	13.33	18.15	21.06	23.68	29.14
16	5.81	7.96	9.31	11.15	12.62	15.33	20.46	23.54	26.29	32.00
18	7.01	9.39	10.86	12.85	14.44	17.33	22.76	25.98	28.86	34.80
20	8.26	10.85	12.44	14.57	16.26	19.33	25.03	28.41	31.41	37.56
22	9.54	12.33	14.04	16.31	18.10	21.33	37.30	30.81	33.92	40.28
24	10.85	13.84	15.65	18.06	19.94	23.33	29.55	33.19	36.41	42.98
26	12.19	15.37	17.29	19.82	21.79	25.33	31.79	35.56	38.88	45.64
28	13.56	16.92	18.93	21.58	23.64	27.33	34.02	37.91	41.33	48.27
30	14.95	18.49	20.59	23.36	25.60	29.33	36.25	40.25	43.77	50.89
40	22.14	26.50	29.06	32.32	34.87	39.33	47.29	51.78	55.75	63.70
50	29.68	34.76	37.70	41.42	44.31	49.33	58.19	63.14	67.50	76.16
60	37.46	43.18	46.47	50.61	53.80	59.33	69.00	74.37	79.08	88.39
100	70.05	77.92	82.38	87.90	92.12	99.33	111.71	118.46	124.34	135.81

9.7 CONFIDENCE LIMITS ON MEDICAL DEVICE MTBF

Generally, in practically inclined reliability studies of medical devices or in that matter of any other engineering items, the times to failure are assumed to follow an exponential distribution. Thus, in testing such items with exponentially distributed times to failure, one can make a point estimate of MTBF. Unfortunately, this provides only an incomplete picture because it does not give any surety of measurement.

It would be more meaningful if we can say, for example, during the testing of a sample of identical medical devices for T_d hours, n number of failures occurred and with a certain degree of confidence the actual MTBF lies somewhere between specific upper and lower limits. Confidence limits may simply be described as the extremes of a confidence interval within which the unknown parameter has a probability of being included.

The lower and upper confidence limits for the test terminated at the preassigned time are as follows, respectively:[6,19]

$$LCL = \frac{2T_t}{\chi^2_{\alpha/2,2n+2}} \tag{9.22}$$

and

$$UCL = \frac{2T_t}{\chi^2_{\left(1-\frac{\alpha}{2}\right),2n}}, \tag{9.23}$$

where

LCL is the lower confidence limit.
UCL is the upper confidence limit.
n is the total number of failures.

Similarly, the lower and upper confidence limits for the test terminated at the preassigned number of failures are as follows, respectively:[6,19]

$$LCL = \frac{2T_t}{\chi^2_{\frac{\alpha}{2},2n}} \tag{9.24}$$

and

$$UCL = \frac{2T_t}{\chi^2_{\left(1-\frac{\alpha}{2}\right),2n}}. \tag{9.25}$$

The values of T_t, when the test is terminated at a preassigned t* and the failed medical devices replaced or nonreplaced, can be calculated by using Equations (9.26) and (9.27), respectively.

$$T_t = Mt^* \tag{9.26}$$

$$T_t = (m-n)t^* + \sum_{i=1}^{n} T_i \tag{9.27}$$

Similarly, in the case of censored medical devices/items (i.e., withdrawal or loss of unfailed items/devices), the values of T_t — when the test is terminated at a preassigned time t^* and the failed medical devices/items replaced but censored items non-replaced or the failed and censored medical devices/items nonreplaced — can be calculated by using Equations (9.28) and (9.29), respectively.

$$T_t = (M - cs)t^* + \sum_{j=1}^{cs} T_j \tag{9.28}$$

and

$$T_t = (M - cs - n)t^* + \sum_{j=1}^{cs} T_j + \sum_{i=1}^{n} T_i , \tag{9.29}$$

where

T_j is the jth, censorship time.
cs is the number of censored medical devices or items.

Example 9.9

A medical device manufacturer put on test 20 identical medical devices at zero time and terminated the test after 500 hours. Five devices failed after 40, 100, 150, 200, and 300 hours of operation, and none of them was replaced. Calculate the medical devices' MTBF and the MTBF's upper and lower limits with a 80% level of confidence.

By inserting the given data into Equation (9.27), we get

$$T_t = (20 - 5)(500) + (40 + 100 + 150 + 200 + 300)$$
$$= 8290 \text{ hours}$$

Thus, the medical devices' MTBF is given by

$$\text{MTBF} = 8290/5$$
$$= 1658 \text{ hours}$$

By substituting, the given and other values into Equations (9.22) and (9.23), we get

$$\text{LCL} = 2(8290)/\chi^2_{0.1,12}$$
$$= 2(8290)/19.54$$
$$= 848.5 \text{ hours}$$

$$\text{UCL} = 2(8290)/\chi^2_{0.9,10}$$
$$= 2(8290)/4.86$$
$$= 3411.5 \text{ hours}$$

It means the medical devices' MTBF is 1658 hours, and the MTBF's upper and lower limits with 80% confidence level are 3411.5 and 848.5 hours, respectively.

9.8 IMPORTANT PUBLICATIONS ON RELIABILITY TESTING

Over the years, many publications have appeared on reliability testing that directly or indirectly relate to medical devices. The important ones are presented below.

- **M1L-HDBK-H108, Sampling Procedures and Tables for Life and Reliability Testing (Based on Exponential Distribution), U.S. Department of Defense, Washington, DC.**
  This handbook provides procedures and tables based on the exponential distribution for life and reliability testing. It includes definitions required for the use of the life test sampling plans and procedures, general description of life test sampling plans, etc.

- **M1L-STD-781D, Reliability Design, Qualification and Production Acceptance Tests: Exponential Distribution, U.S. Department of Defense, Washington, DC.**
  This document covers the requirements and provides details for reliability testing during the development, qualification, and production of systems and equipment with an exponential time to failure distribution. Test time is stated in multiples of the design mean time between failures (MTBF). Specifying any two of three parameters, i.e., lower test MTBF, upper test MTBF, or their ratio, given the desired decision risks, determines the test plan to be used.

- **MIL-STD-2074, Failure Classification for Reliability Testing, U.S. Department of Defense, Washington, DC.**
  This document establishes criteria for classification of failures occurring during reliability testing. The classification into relevant or nonrelevant categories allows the proper generation of MTBF reports. The document applies to any reliability test, including, but not restricted to, tests performed in accordance with M1L-STD-781.

- **M1L-HDBK-781, Reliability Test Methods, Plans, and Environments for Engineering Development, Qualification and Production, U.S. Department of Defense, Washington, DC.**
  This handbook provides test methods, test plans, and test environmental profiles that can be used in reliability testing during the development, qualification, and production of systems and equipment. The document is designed to be used with M1L-STD-781.

- **M1L-HDBK-728, Non-Destructive Testing (NDT), U.S. Department of Defense, Washington, DC.**
This document serves as a guide and describes general principles, procedures and safety items, of eddy current, liquid penetrate, magnetic particle, radiographic and ultrasonic testing. The handbook can serve as a ready reference to the important principles and facts relating to the employment of nondestructive testing, inspection and evaluation.

- **M1L-STD-690C, Failure Rate Sampling Plans and Procedures, U.S. Department of Defense, Washington, DC.**
This document provides procedures for failure rate qualification, sampling plans for establishing and maintaining failure rate levels at selected confidence levels, and a lot conformance inspection procedures associated with failure rate testing for the purpose of direct reference in appropriate military electronic parts established reliability (ER) specifications. Figures and tables throughout this document are based on exponential distribution.

- **M1L-STD-202F, Test Methods for Electronic and Electrical Component Parts, U.S. Department of Defense, Washington, DC.**
This document establishes uniform methods for testing electronic and electrical component parts, including basic environmental tests to determine resistance to deleterious effects of natural elements and conditions surrounding military operations, and physical and electrical tests.

- **M1L-STD-2165, Testability Program for Electronic Systems and Equipment, U.S. Department of Defense, Washington, DC.**
This military document defines methodology for incorporating adequate and cost-effective testability and built-in test (BIT) features into the equipment design. It sets the requirements and establishes guidelines for assessing the extent to which a system or a unit supports fault detection and fault isolation. The document addresses three types of tasks: program monitoring and control tasks, design and analysis tasks, and test and evaluation tasks.

- **M1L-STD-810E, Environmental Test Methods and Engineering Guidelines, U.S. Department of Defense, Washington, DC.**
The purpose of this military standard is to standardize the design and conduct of tests for assessing the ability of equipment to withstand environmental stresses it will encounter during its life cycle and to make sure plans and test results are adequately documented.

- **Nash, F.R., *Estimating Device Reliability: Assessment of Credibility*, Kluwer Academic Publisher, Boston, 1993.**
This book indirectly relates to the reliability testing of medical devices.

- **IEC 1123, Reliability Testing Compliance Test Plans for Success Ratio, International Electrotechnical Commission (IEC), Switzerland.**
- **BSI BS 5760 (Sec. 10), Reliability of Systems, Equipment and Components Part 10, Guide to Reliability Testing Section, General Requirements, Compliance Test Procedures for Steady State Availability, Guide to Reliability Testing, Compliance Test Plans for Success Ratio, British Standards Institute London, U.K.**
- **IEC 605, Equipment Reliability Testing, International Electrotechnical Commission, Switzerland.**
- **MOD UK DSTAN 00-43, Reliability and Maintainability Assurance Activity Part I: In-service Reliability Demonstrations Issue, British Defense Standards, London, U.K.**

9.9 PROBLEMS

1. Discuss the following types of testing:
   - Stress testing
   - Event testing
2. Define the term "reliability testing."
3. What are the classifications of reliability tests?
4. What is accelerated testing?
5. What are the ways to perform accelerated testing?
6. A manufacturer of medical devices tested 20 identical devices to failure at 120°C, and their MTBF was 2000 hours. Calculate the MTBF of the medical devices at the normal operating temperature of 85°C.
7. A manufacturer of certain medical items is required to demonstrate 95% reliability of items it manufactures at 90% confidence level. Determine the total number of items to be tested if no failures are allowed.
8. Describe the following terms:
   - Thermal shock testing
   - Impact testing
   - Success testing
9. Twenty-five identical medical devices were tested for 3000 hours, and none of them failed. Calculate the lower limit of the device MTBF at the 80% confidence level.
10. A company manufacturing medical devices placed on test 40 identical devices and tested them for 1000 hours. Three devices failed after 100, 200, and 400 hours of operation and none of them was replaced. Calculate the medical devices' MTBF and the MTBF's upper and lower limits with 90% confidence level.

REFERENCES

1. MIL-STD-781, Reliability Design Qualification and Production Acceptance Test: Exponential Distribution, U.S. Department of Defense, Washington, DC.
2. MIL-HDBK-781, Reliability Test Methods, Plans and Environments for Engineering Development, Qualification and Production, U.S. Department of Defense, Washington, DC.
3. Dhillon, B.S., *Reliability and Quality Control: Bibliography on General and Specialized Areas*, Beta Publishers, Inc., Gloucester, Ontario, 1992.
4. Fries, R.C., *Reliable Design of Medical Devices*, Marcel Dekker Inc., New York, 1997.
5. Fries, R.C., *Reliability Assurance for Medical Devices, Equipment and Software*, Interpharm Press Inc., Buffalo Grove, IL, 1991.
6. Von Alven, W.H., Ed., *Reliability Engineering*, Prentice-Hall Inc., Englewood Cliffs, NJ, 1964.
7. Nelson, W., *Accelerated Testing*, John Wiley & Sons Inc., New York, 1980.
8. Meeker, W.Q. and Hahn, G.J., *How to Plan an Accelerated Life Test: Some Practical Guidelines*, American Society for Quality Control (ASQC), Milwaukee, WI, 1985.
9. Bain, L.J. and Engelhardt, M., *Statistical Analysis of Reliability and Life Testing Models: Theory*, Marcel Dekker Inc., New York, 1991.
10. Dhillon, B.S., *Design Reliability: Fundamentals and Applications*, CRC Press, Boca Raton, FL, 1999.
11. Jensen, F. and Peterson, N.E., Burn-In, John Wiley & Sons Inc., New York, 1982.
12. Tobias, P.A., Trindade, D., *Applied Reliability*, Van Nostrand Reinhold Company, New York, 1986.
13. Elsayed, A.E., *Reliability Engineering*, Addison Wesley Longman Inc., Reading, MA, 1996.
14. Nelson, W., A Survey of Methods for Planning and Analysing Accelerated Life Tests, *IEEE Trans. Electr. Insul.*, Vol. 9, 1974, pp. 12–18.
15. Yurkowsky, W., Schafer, R.E., and Finkelstein, J.M., Accelerated Testing Technology, Report No. RADC-TR-67-420, Rome Air Development Center, Griffiss Air Force Base, Rome, NY, 1967.
16. Grant Ireson, W., Coombs, C.F., and Moss, R.Y., *Handbook of Reliability Engineering and Management*, McGraw-Hill, New York, 1996.
17. Dhillon, B.S., *Design Reliability: Fundamentals and Applications*, CRC Press, Boca Raton, FL, 1999.
18. AMC Pamphlet 702-3, U.S. Army Material Command (AMC), Headquarters, Washington, DC, 1968.

10 Medical Device Costing

CONTENTS

10.1 INTRODUCTION

Cost is an important factor in health care, and over the years it has been increasing at an alarming rate. For example, the total expenditures on health care in the U.S. have increased over 800% from 1960 to 1986, and the projection for the year 2000 is around $1.5 trillion.[1] Data available for 1986 indicate that hospital care accounted for the largest share of the total health care expenditures of $465 billion or 39%.

A significant amount of hospital care expenditures is spent on medical devices and equipment. Just as in the case of any other engineering equipment, the cost estimating in medical devices is important to make an effective decision. A medical device or a piece of equipment designed, developed, or maintained without paying any proper attention to the cost consideration could be a costly affair due to various factors. The important components of the total cost of producing a medical device are design and development cost, materials cost, testing cost, production cost, and maintenance cost. In particular, recent data indicate that the maintenance cost of software-driven medical products is high and is approximately 50% of the cost of a software product. Maintenance and software support functions include actions such as analyzing and removing errors and providing enhancements.[2]

This chapter presents various aspects of medical device costing.

10.2 REASONS FOR MEDICAL DEVICE COSTING AND MEDICAL DEVICE DEVELOPMENT ECONOMIC ANALYSIS FACTORS

There could be many reasons for medical device costing; some are as follows:[3-4]

- To ascertain the most profitable procedure and materials during manufacture of the device.
- To determine the amount to be spent on equipment and other related items to manufacture the device.
- To determine the device's selling price.
- To determine if it is cheaper to manufacture the parts in-house or to purchase them from vendors.
- To provide input to the long-term financial goals of the company.
- To conduct new device feasibility analysis.
- To provide assistance in controlling the cost of device manufacturing.
- To determine the efficiency of the device manufacturing process.

There are many economic analysis factors associated with medical device development. The economic analysis should include factors such as those shown in Figure 10.1.[2]

10.3 METHODS FOR MAKING MEDICAL DEVICE INVESTMENT DECISIONS

The selection of medical device development projects is important to organizations involved with medical devices. With respect to the financial aspect, it could be the

FIGURE 10.1 Factors relating to medical device economic analysis.

factor behind the success or failure of an organization. Fortunately, over the years many methods and techniques have been developed to select product development projects for investment. Such approaches can also be used for making medical device investment decisions. This section presents some of these methods.[4-8]

10.3.1 Payback Period Method

This involves determining the payback time over which the project capital expenditure can be recovered. The project is considered attractive if its useful life is much greater than the payback period. And if its useful life is equal to or a little more than the payback time, it is considered unattractive.

The main drawback of this model is that it does not consider the profit factor once the capital expenditures have been recovered. The method uses four approaches to calculate payback period.[4,8]

Approach I

The payback period is defined by

$$T_I = \frac{CE}{P_g}, \tag{10.1}$$

where

T_I is the payback period or time for approach I.
CE is the capital expenditure.
P_g is average annual gross profit.

Approach II

The payback time is expressed by

$$T_{II} = \frac{CE}{P_n}, \tag{10.2}$$

where

T_{II} is the payback time or period for approach II.
P_n is the average annual net profit.

In turn, the net profit, P_n, is defined by

$$P_n = P_g - \left[TR\left(P_g - (DR)(CE)\right) + i(CE + WC)\right], \tag{10.3}$$

where

TR is the income tax rate.
DR is the invested capital (fixed) depreciation rate.
i is the interest rate on the borrowed money.
WC is the working capital.

Approach III

The payback period is given by

$$T_{III} = \frac{CE}{CF}, \tag{10.4}$$

where

T_{III} is the payback period or time for approach III.
CF is the average annual cash flow.

The average annual cash flow is given by

$$CF = P_n + (CE)\,(DR). \tag{10.5}$$

Approach IV

The payback period is expressed by

$$T_{IV} = \left[DR + \{\alpha/(1 - TR)\}\right]^{-1}, \tag{10.6}$$

where

T_{IV} is the payback period or time for approach IV.
α is the minimum acceptable rate of return on investment.

Example 10.1

A health care facility wishes to replace a piece of equipment at a cost of $100,000. Its use will generate an average annual saving of $8000. Determine the capital expenditure recovery period.

Substituting the given data into Equation (10.1) yields

$$T_I = \frac{100,000}{8000} = 12.5 \text{ years}$$

It means the capital expenditure recovery period will be 12.5 years.

10.3.2 Return on Investment Method

This is another method that can be used for making medical device investment decisions. Compared with the payback period technique, this method is relatively more complex. It uses two approaches to compute the return on investment.

Approach I

In this case, the return on original investment is defined by

$$R_O = \frac{NP_m(100)}{(AM + WC)}, \tag{10.7}$$

where

R_O is the return on original investment.
NP_m is the annual average net profit.
AM is the amount invested.

Approach II

In this case, the return on average investment is given by

$$R_a = \frac{NP_m(100)}{(0.5\,AM + WC)}, \tag{10.8}$$

where

R_a is the return on average investment.

Example 10.2

A hospital is considering buying a medical system that requires $250,000 in investment and $10,000 in working capital. It is estimated that the use of the system will generate average net savings or profit of $20,000 per year. Calculate the return on original investment.

By substituting the given data into Equation (10.7), we get

$$R_O = \frac{20,000(100)}{(250,000)+10,000} = 7.69\%.$$

Thus, the return on the original investment will be 7.69%.

10.3.3 Benefit/Cost Analysis Method

This is a widely used method, and it is used to determine if benefits from the project under consideration outweigh its costs. More specifically, the project is considered for development only if its benefits are greater than the investment cost. The benefit/cost ratio is defined by

$$\mathrm{BCR} = \frac{\mathrm{UB}}{\mathrm{TIC}}, \tag{10.9}$$

where

BCR is the benefit-cost ratio.
UB is the user benefits.
TIC is the total investment cost including the operating cost.

Example 10.3

A piece of medical equipment is being considered for procurement, and its cost and useful life are $150,000 and ten years, respectively. The annual maintenance cost of the equipment is estimated to be $1500. The total benefits of the equipment over its life span are estimated to be $300,000. Calculate the benefit/cost ratio.

The total investment cost of the medical equipment under consideration is

$$\begin{aligned}\mathrm{TIC} &= 150,000 + (1500)(10)\\ &= \$165,000.\end{aligned}$$

Inserting the above calculated value and the other given data into Equation (10.9) yields

$$\mathrm{BCR} = \frac{300,000}{165,000} = 1.82.$$

Thus, the value of the benefit-cost ratio is 1.82.

10.4 FORMULAS FOR DETERMINING THE TIME VALUE OF MONEY

With the passage of time, the value of money changes. Consequently, the cost of an item is affected. Therefore, this section presents some useful formulas for determining the time value of money in medical device costing.[9]

10.4.1 Single Payment Present Value Formula

This formula is used to determine the present value of a single sum of money to be received or spent after *n* interest periods or years. Thus, the present value is given by

$$PV = F(1 + i)^{-n}, \tag{10.10}$$

where

PV is the present value of the future amount F.
n is the number of interest periods or years.
i is the interest rate per interest period or year.

Example 10.4

Assume that the salvage value of an X-ray machine after its useful life period of ten years will be $50,000. If the annual average interest rate is 5%, calculate the present value of the X-ray machine salvage value. By inserting the specified data into Equation (10.10), we get

$$PV = (50,000)(1+0.05)^{-10}$$
$$= \$30,695.70$$

Thus, the present value of the X-ray machine is $30,695.70.

10.4.2 Uniform Periodic Payment Present Worth Formula

This formula determines the present worth of equal payments made at the end of each of *n* interest periods or years. All these payments are deposited at a compound interest rate *i*. Thus, the present worth, *PW*, of all these periodic payments is given by

$$PW = PW_1 + PW_2 + \ldots + PW_n$$
$$= \frac{Z}{(1+i)} + \frac{Z}{(1+i)^2} + \ldots + \frac{Z}{(1+i)^n}, \tag{10.11}$$

where

Z is the amount of payment.
PW_j is the present worth of payment Z made at the end of interest period or year j; for j = 1, 2, …, n.

Equation (10.11) is a geometric series, thus after finding its sum, we write it as follows:

$$PW = Z\left[\frac{1-(1+i)^{-n}}{i}\right]. \tag{10.12}$$

Example 10.5

Assume that the expected useful life of an ultrasound device is 15 years and its estimated annual maintenance cost is \$1000. Determine the present worth of the total maintenance cost if the annual interest rate is estimated to be 7%.

Inserting the given data into Equation (10.12) yields

$$PW = 1000\left[\frac{1-(1+0.07)^{-15}}{0.07}\right]$$

$$= \$9,107.90$$

It means the present value of the total maintenance cost of the ultrasound device is \$9,107.90.

10.5 USEFUL COST-ESTIMATION MODELS FOR MEDICAL DEVICES

Over the years, many cost-estimation models have been developed. This section presents some of those models useful for performing directly or indirectly cost-related reliability studies of medical devices.[10,11]

10.5.1 Failure Mode and Effect Analysis (FMEA) Cost-Estimation Model

The cost of performing FMEA is expressed by

$$C_{fm} = (C_{mhf})\,(MH_f), \tag{10.13}$$

where

C_{fm} is the cost of performing FMEA.
C_{mhf} is the cost per hour of performing FMEA.
MH_f is the total number of hours required for performing FMEA.

In turn, the total number of hours is given by

$$MH_f = (17.79)n, \tag{10.14}$$

where

n is the number of unique items requiring FMEA, the number of pieces of equipment for equipment level FMEA, or the number of circuit cards for piece part and circuit level FMEAs. Its value varies from 3 to 206.

10.5.2 Reliability Modelling/Allocation Cost-Estimation Model

The cost of reliability modelling/allocation is defined by

$$C_{m/a} = (C_{mhr})\,(MH_r), \tag{10.15}$$

where

$C_{m/a}$ is the cost of reliability modelling/allocation.
C_{mhr} is the cost per hour of performing reliability modelling/allocation.
MH_r is the total number of hours required for performing reliability modelling/allocation.

The total number of hours, MH_r, is given by

$$MH_r = (4.05)\,\alpha^2\beta, \tag{10.16}$$

where

α is the variable whose value depends on the degree of modelling and allocation complexity; series system ($\alpha = 1$), simple redundancy ($\alpha = 2$), and very complex redundancy ($\alpha = 3$).
β is the number of items in the allocation process; its value varies from 7 to 445.

10.5.3 Reliability Testing Cost-Estimation Model

The cost of reliability testing is expressed by

$$C_{rt} = (C_{mht})\,(MH_t), \tag{10.17}$$

where

C_{rt} is the cost of reliability testing.
C_{mht} is the cost per hour of conducting reliability testing.
MH_t is the total number of hours required for performing reliability testing.

The total number of hours, MH_t, can be determined by using the following equation:

$$MH_t = (182.07)k, \tag{10.18}$$

where

k is the factor whose value depends on hardware complexity; for less than 15,000 parts (k = 1), for parts between 15,000 and 25,000 (k = 2), and for parts greater than 25,000 (k = 3).

10.5.4 Reliability Prediction Cost-Estimation Model

The cost of reliability prediction is

$$C_{rp} = (C_{mhp})(MH_p), \tag{10.19}$$

where

C_{rp} is the cost of reliability prediction.
C_{mhp} is the cost per hour of performing reliability prediction.
MH_p is the total number of hours needed for performing reliability prediction.

The total number of hours, MH_p, is given by

$$MH_p = (4.54)F_1^2 F_2^2 F_3, \tag{10.20}$$

where

F_1 is the factor that takes into consideration the level of detail; prediction exists ($F_1 = 1$), prediction made using similar system data ($F_1 = 2$), and full MIL-HDBK-217[12] stress prediction ($F_1 = 3$),
F_2 is the factor that takes into consideration the report formality; internal report ($F_2 = 1$), and formal report required ($F_2 = 2$).
F_3 is the factor that takes into consideration the percentage of commercial hardware used; 0–25% ($F_3 = 4$), 26–50% ($F_3 = 3$), 51–75% ($F_3 = 2$), and 76–100% ($F_3 = 1$).

10.5.5 Failure Reporting and Corrective Action System (FRACAS) Cost-Estimation Model

The cost of developing FRACAS is expressed by

$$C_{fs} = (C_{mhs})(MH_s), \tag{10.21}$$

where

C_{fs} is the cost of developing FRACAS.
C_{mhs} is the cost per hour associated with developing FRACAS.
MH_s is the total number of hours required for developing FRACAS.

In turn, the total number of hours, MH_s, is given by

$$MH_s = (8.25)D^2, \tag{10.22}$$

where

D is the duration of FRACAS implementation expressed in months; it varies from 2.5 to 38.

10.5.6 Reliability and Maintainability (R and M) Program Plan Cost-Estimation Model

The cost of developing an R and M program plan is expressed by

$$C_{rm} = (C_{mhm})\,(MH_m), \tag{10.23}$$

where

C_{rm} is the cost of developing the R and M program plan.
C_{mhm} is the cost per hour of developing the R and M program plan.
MH_m is the total number of hours needed to develop the R and M program plan.

In this case, the total number of hours, MH_m, is given by

$$MH_m = (2.73)x^2, \tag{10.24}$$

where

x is the number of MIL-STD-785/470[13,14] tasks required; its value varies from 4 to 22.

10.5.7 Corrective Maintenance Cost-Estimation Model

This model is used to estimate yearly labor cost of corrective maintenance associated with an item. Thus, the corrective maintenance cost is expressed by

$$CMC = (CMCH)\,(SOH)\,(MTTR)/MTBF, \tag{10.25}$$

where

CMC is the annual corrective maintenance labor cost.
$CMCH$ is the hourly corrective maintenance labor cost.
SOH is the scheduled operating hours of the item per year.
$MTBF$ is the mean time between failures of the item.
$MTTR$ is the mean time to repair of the item.

Example 10.6

Assume that a medical equipment is scheduled for operation 5000 hours annually, and its estimated MTBF and MTTR are 1000 hours and 10 hours, respectively. Calculate the annual corrective maintenance labour cost of the medical equipment if the hourly maintenance labor rate is $25.

By inserting the given data into Equation (10.24), we get

$$\text{CMC} = (25)(5000)(10)/(1000)$$

$$= \$1250 .$$

Thus, the annual corrective maintenance labor cost of the medical equipment is $1250.

10.5.8 Software Maintenance Cost-Estimation Model

The following equation can be used to estimate medical software maintenance cost:[15]

$$\text{SMC} = 3(\text{CMM})\theta_1/\theta_2 \tag{10.26}$$

where

SMC is the software maintenance cost.
θ_1 is the number of instructions to be changed per month.
θ_2 is the difficulty constant.
CMM is the cost per month.

10.5.9 Software Development Cost-Estimation Model

This model can be used to estimate the medical device software development cost. The total software development cost is expressed by.[10,16,17]

$$\text{SDC} = \text{PC} + \text{SC}, \tag{10.27}$$

where

SDC is the total software development cost.
PC is the primary cost of the software development that basically includes the cost of labour.
SC is the secondary cost of the software development. Some examples of the components of the secondary cost are cost of documentation and cost of computer time.

The primary cost, PC, of developing software is expressed by

$$\text{PC} = \text{R}_a(\text{MM}), \tag{10.28}$$

where

R_a is the average labor rate associated with the software development manpower expressed in dollars per month. It includes administration cost, overhead cost, general cost, fees, and so on.
MM is the manpower required for developing the software expressed in months. The elements of the software development are design, analysis, code, debug, test, and checkout.

The secondary cost of the software development is expressed by

$$SC = \sum_{i=1}^{m} RC_{si} = \mu(PC) = \mu R_a MM, \qquad (10.29)$$

where

- m is the number of secondary resources.
- RC_{si} is the secondary resource i cost; for $i = 1, 2, 3, \ldots, m$.
- μ is the ratio of secondary software development cost to primary software development cost.

10.5.10 Cost-Capacity Model

This is a widely used model to approximate cost of items of different capacity when data on similar item cost and capacity are available. Thus, the cost of the new item is expressed by[18,19]

$$C_n = C_s \left[\frac{KP_n}{KP_s}\right]^{\theta}, \qquad (10.30)$$

where

- C_n is the cost of the new item.
- KP_n is the capacity of the new item.
- C_s is the known cost of the similar item of capacity KP_s.
- θ is the cost-capacity factor whose value varies for different items.[10,20] However, in the event of having no data for the factor, it is reasonable to assume 0.6 as its value.

Example 10.7

Assume that the cost of a medical facility that can handle ten patients per day is \$0.8 million. Calculate the cost of a new medical facility that can handle 20 patients per day if the value of the cost-capacity factor is 0.7.

Substituting the given values into Equation (10.30) yields

$$C_u = (0.8)\left[\frac{20}{10}\right]^{0.7} = \$1.2996 \text{ million}.$$

Thus, the cost estimate for the new medical facility is \$1.2996 million.

10.6 LIFE CYCLE COSTING

The history of the life cycle costing concept may be traced back to 1965 when the Logistics Management Institute of Washington, DC published a document entitled "Life Cycle Costing in Equipment Procurement."[21] The life cycle cost (LCC) of an item may simply be described as the sum of all costs, i.e., procurement and ownership, over its entire life span. The application of the cycle costing concept in developing medical devices could be effective in performing cost-effectiveness analysis. Prior to starting a life cycle costing study of an item, it is essential to seek answers to questions on diverse areas concerning that item. These areas include required data, goal of the estimate, ground rules and assumptions, fund limitations, analysis constraints, time schedule, estimating procedures, treatment of uncertainties, involved personnel, cost analyst responsibility, required analysis details, analysis format, life cycle costing group, users of the end results, analysis required precision and accuracy, and life cycle costing process auditing and controlling.[10,15,22]

10.6.1 Life Cycle Costing Procedures

The life cycle costing procedure is composed of many steps, as shown in Figure 10.2.[23]

10.6.2 Life Cycle Cost Model Inputs

There are many inputs to a life cycle cost model, and they range from the cost of labor per corrective maintenance action to the cost of training. Some of those inputs are item mean time between failures and mean time to repair, cost of installation, procurement cost, cost of training, time spent for travel, warranty coverage period, requirement for spares, cost of labor per preventive maintenance action, average material cost of a failure, and cost of carrying spares in inventory.[24]

10.6.3 Life Cycle Cost Models

The life cycle cost models developed over the years may be grouped under two classifications: general and specific. The general life cycle cost models are not tied to any specific item. On the other hand, the specific life cycle cost models are tied to specific items, for example, electric motor, tank gun system, and early warning radar.

One general and one specific life cycle cost models are presented below.

General Life Cycle Cost Model

In this case, the life cycle cost (LCC) of an item is composed of two major components: recurring cost and nonrecurring cost. Mathematically, it is expressed by[10,25]

$$LCC = RC + NRC, \tag{10.31}$$

where

RC is the recurring cost of the item.
NRC is the nonrecurring cost of the item.

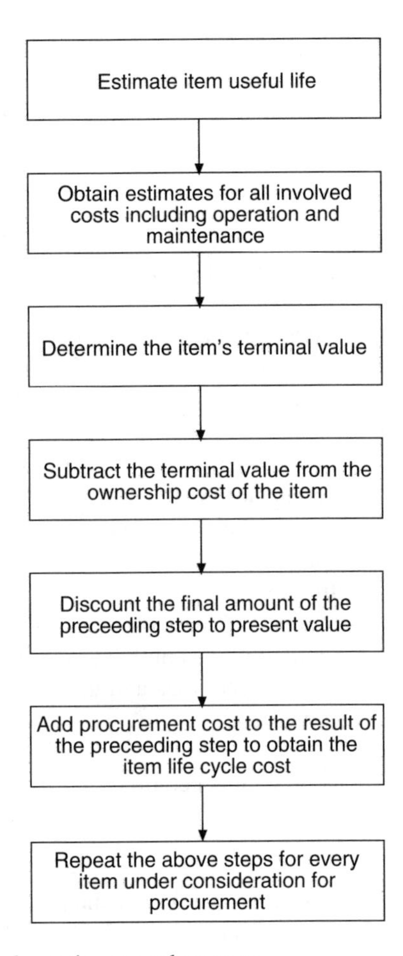

FIGURE 10.2 Life cycle costing procedure steps.

The elements of the recurring cost include maintenance cost, manpower cost, support cost, operating cost, and inventory cost. Similarly, the important elements of the nonrecurring cost are procurement cost, installation cost, training cost, research and development cost, support cost, qualification approval cost, reliability and maintainability improvement cost, cost of life cycle management, cost of test equipment, and transportation cost.

Specific Model: Health Care Facility Life Cycle Cost Model

The life cycle cost of a health care facility is expressed by[10,26]

$$LCC_h = C_p + C_o, \tag{10.32}$$

where

- LCC_h is the health care facility life cycle cost.
- C_p is the health care facility capital cost.
- C_o is the health care facility operating cost.

The components of the capital cost, C_p, are land acquisition cost, financing cost, collateral equipment cost, demolition and site preparation cost, indirect cost, direct construction or purchase cost, denial of use cost, and alteration and replacement cost.

The elements of the operating cost, C_o, are as follows:

- Utility and fuel cost
- Painting cost
- Equipment (furnishing) maintenance cost
- Structural maintenance cost
- Heating system operations and maintenance cost
- Incinerator and trash removal cost
- Fire protection system maintenance cost
- Space changes cost
- Equipment (fixed equipment and specific construction) maintenance cost
- Grounds and roads maintenance cost
- Elevator, escalator, and dumbwaiter operations cost
- Exterior building cleaning cost
- Insect and rodent control cost
- Electrical system operations and maintenance cost
- Air conditioning and ventilating systems operations and maintenance cost
- Special mechanical systems operations and maintenance cost
- Internal buildings cleaning cost
- Exterior restoration cost
- Plumbing and sewerage systems operations and maintenance cost

Example 10.8

A hospital is considering procuring an X-ray machine, and two manufacturers are bidding to sell the equipment. The data given in Table 10.1 are available on both manufacturers' machines. Determine which of the two X-ray machines would the hospital buy with respect to their life cycle costs.

TABLE 10.1
Data for X-ray Machines Under Consideration

No.	Description	Manufacturer A's X-ray Machine	Manufacturer B's X-ray Machine
1	Price	$180,000	$160,000
2	Annual interest rate (cost of money)	6%	6%
3	Annual failure rate	0.04 failures	0.03 failures
4	Useful operating life	15 years	15 years
5	Expected cost of a failure	$6000	$5000
6	Annual operating cost	$2000	$5000

Manufacturer A's X-ray Machine

The expected cost, EC_a, of a failure is

$$EC_a = (0.04)(6000)$$
$$= \$240 .$$

By substituting the above calculated value and the given data into Equation (10.12), we get the following present value of the life cycle maintenance cost

$$PV_{am} = 240\left[\frac{1-(1+0.06)^{-15}}{0.06}\right]$$
$$= \$2,330.9 ,$$

where

PV_{am} is the present value of the life cycle maintenance cost of manufacturer A's X-ray machine.

Similarly, the present value, PV_{ao}, of the life cycle operating cost of manufacturer B's X-ray machine is

$$PV_{ao} = 2000\left[\frac{1-(1+0.06)^{-15}}{0.06}\right]$$
$$= \$19,424.5 .$$

Using the above calculated values and the given data, we get the following life cycle cost, LCC_a, of manufacturer A's X-ray machine

$$LCC_a = 180,000 + 2330.9 + 19,424.5$$
$$= \$201,755.4$$

Manufacturer B's X-ray Machine

The expected cost, EC_b, of a failure is

$$EC_b = (0.03)\ (5,000) = \$150$$

By inserting the above calculated value and the given data into Equation (10.12), we get the following present value of the life cycle maintenance cost

$$PV_{bm} = 150\left[\frac{1-(1+0.06)^{-15}}{0.06}\right]$$

$$= \$1456.8,$$

where

PV_{bm} is the present value of the life cycle maintenance cost of manufacturer B's X- ray machine.

Similarly, the present value, PV_{bo}, of the life cycle operating cost of manufacturer B's X-ray machine is

$$PV_{bo} = 5000\left[\frac{1-(1+0.06)^{-15}}{0.06}\right]$$

$$= \$48,561.2.$$

By using the above calculated values and the given data, we get the following life cycle cost, LCC_b, of manufacturers B's X-ray machine

$$LCC_b = 160,000 + 1456.80 + 48,561.2$$

$$= \$210,018.$$

It means, it would be cheaper for the hospital to buy manufacturer A's X-ray machine with respect to life cycle cost.

10.7 PROBLEMS

1. List the reasons for medical device costing.
2. What are the economic analysis factors associated with the medical device development?
3. Describe the following two methods associated with making medical device investment decisions:
   - Payback period method
   - Return on investment method
4. Assume that the terminal value of an ultrasound equipment after its useful life of 15 years is $30,000. Determine the present value of the equipment terminal value by assuming the average annual interest rate of 6%.
5. It is estimated that the useful life and the annual maintenance cost of a piece of medical equipment are 10 years and $1500, respectively. Calculate the present value of the total maintenance cost if the annual interest rate is expected to be 8%.

6. Assume that an X-ray machine is scheduled for operation, 3000 hours annually, and its expected MTBF and MTTR are 500 hours and 5 hours, respectively. Determine the annual corrective maintenance labor cost of the X-ray machine if the hourly maintenance labor rate is $30.
7. Define the term "life cycle cost."
8. Write an essay on the historical development of the life cycle costing concept.
9. Describe the life cycle costing procedure.
10. A health care facility is considering buying a medical equipment, and three manufacturers are bidding to sell the equipment. Table 10.2 presents data available on all manufacturers' medical equipment. Determine, which one of the three medical equipments would the health care facility buy with respect to their life cycle costs.

TABLE 10.2
Data for Medical Equipment Under Consideration

No.	Description	Manufacturer A's Medical Equipment	Manufacturer B's Medical Equipment	Manufacturer C's Medical Equipment
1	Price	$200,000	$220,000	$240,000
2	Annual failure rate	0.02 failures	0.03 failures	0.04 failures
3	Annual interest rate (cost of money)	7%	7%	7%
4	Annual operating cost	$6000	$4500	$3000
5	Expected cost of a failure	$5000	$6000	$7000
6	Annual operating life	10 years	10 years	10 years

REFERENCES

1. Rice, J.A. and Garside, P.A., The Impact of Cost Control in Restructuring of the Health-care Industry, *Medical Device Industry*, N.F. Estrin, Ed., Marcel Dekker, Inc., New York, 1990, pp. 703-718.
2. Fries, R.C., *Reliable Design of Medical Devices*, Marcel Dekker, Inc., New York, 1997.
3. Doyle, L.E., How to Estimate Costs of New Products, in *Management Guide for Engineers and Technical Administrators*, Chironis, N.P., Ed., McGraw-Hill Book Company, New York, 1969.
4. Dhillon, B.S., *Engineering Management*, Technomic Publishing Company, Lancaster, PA, 1987.
5. Cetron, M.J., Martino, J., and Roepcke, L., The Selection of R and D Program Content: Survey of Quantitative Methods, *IEEE Transactions on Engineering Management*, Vol. 14, 1967, pp. 4–13.
6. Baker, N.R. and Pound, W.J., R and D Project Selection: Where We Stand, *IEEE Transactions on Engineering Management*, Vol. 11, 1964, pp. 124–134.

7. Clarke, T.E., Decision Making in Technologically Based Organizations: A Literature Survey of Present Practice, *IEEE Transactions on Engineering Management*, Vol. 21, 1974, pp. 9–23.
8. Kasner, E., *Essentials of Engineering Economics*, McGraw-Hill Book Company, New York, 1979.
9. White, J.A., Agee, M.H., and Case, K.E., *Principles of Engineering Economic Analysis*, John Wiley & Sons, Inc., New York, 1977.
10. Dhillon, B.S., *Life Cycle Costing*, Gordon and Breach Science Publishers, New York, 1989.
11. RADC Reliability Engineer's Toolkit, prepared by the Systems Reliability and Engineering Division, Rome Air Development Center (RADC), Griffiss Air Force Base, Rome, NY 13441-5700, 1988.
12. MIL-HDBK-217, Reliability Prediction of Electronic Equipment, Department of Defense, Washington, DC.
13. MIL-STD-785, Reliability Program for Systems and Equipment Development and Production, Department of Defense, Washington, DC.
14. MIL-STD-470, Maintainability Program for Systems and Equipment, Department of Defense, Washington, DC.
15. Sheldon, M.R., *Life Cycle Costing: A Better Method of Government Procurement*, Westview Press, Boulder, CO, 1979.
16. Herd, J.H., Postak, J.N., Russell, W.E., and Stewart, K.R., *Software Cost Estimation Study*, Vol. I, Report No. RADC-TR-77-220, 1977. Available from Doty Associates, Inc., 416 Hungerford Drive, Rockville, MD 20850, USA.
17. Doty, D.L., Nelson, P.J., and Stewart, K.R., Software Cost Estimation Study, Vol. II, Report No. RADC-TR-77-220, 1977. Available from Doty Associates, Inc., 416 Hungerford Drive, Rockville, MD 20850, USA.
18. Dieter, G.E., *Engineering Design*, McGraw-Hill Book Company, New York, 1983.
19. Dhillon, B.S., *Engineering Design: A Modern Approach*, Richard D. Irwin, Inc., Chicago, 1996.
20. Desai, M.B., Preliminary Cost Estimating of Process Plants, *Chemical Engineering*, July 1981, pp. 65–70.
21. Report No. LMI Task 4C-5, Life Cycle Costing in Equipment Procurement, Prepared by the Logistics Management Institute (LMI), Washington, DC, April 1965.
22. Dhillon, B.S. and Reiche, H., *Reliability and Maintainability Management*, Van Nostrand Reinhold Company, New York, 1985.
23. Coe, C.K., Life Cycle Costing by State Governments, Public Administration Review, September/October 1981, pp. 564–569.
24. Siewiorek, D.P. and Swarz, R.S., The Theory and Practice of Reliable System Design, Digital Press, Digital Equipment Corporation, Bedford, MA, 1982.
25. Reiche, H., Life Cycle Cost, in *Reliability and Maintainability of Electronic Systems*, J.E. Arsenault and J.A. Roberts, Eds., Computer Science Press, Potomac, MD, 1980, pp. 3–23.
26. Eddins-Earles, M., *Factors, Formulas, and Structures for Life Cycle Costing*, Published by Eddins-Earles, 89 Lee Drive, Concord, MA, 1981.

11 Medical Device Maintenance and Maintainability

CONTENTS

11.1 INTRODUCTION

The maintenance of engineering equipment is as important as the equipment's design and development. Usually, much more money is spent on maintaining a piece of equipment over its life span than on its procurement. For example, a study conducted by the U.S. Air Force (USAF) over a period of five years revealed that repair and maintenance costs were approximately ten times the original cost.[1]

In modern times, concerns regarding the consideration of maintenance aspects during the design may be traced back to the development of the Wright brothers' airplane in 1901.[2] The contract signed between the Army Signal Corps and Wright brothers specifically stated that the airplane be "simple to operate and maintain." This led us to equipment maintainability, which may simply be described as a characteristic of product/system design. It concerns system/product attributes such as controls, displays, accessibility, test equipment, and test points. In contrast, maintenance may be stated as all actions appropriate for retaining an equipment or item in, or restoring to, a required condition.[3]

Just like in the case of general engineering equipment, maintenance and maintainability are important factors in medical devices/equipment. Some of the important publications on medical device/equipment maintenance/maintainability are listed in references 4–7.

This chapter discusses the important aspects of maintenance and maintainability directly or indirectly related to medical devices or equipment.

11.2 TERMS AND DEFINITIONS

This section presents selective terms and definitions associated with maintenance and maintainability considered useful for medical devices/equipment:[3,8-11]

- **Maintenance**. This is all actions appropriate for retaining and equipment/item in, or restoring to, a given condition.
- **Maintainability**. This is the probability that a failed piece of equipment or item will be restored to its acceptable operational condition.
- **Corrective maintenance**. These are actions taken because of a failure to restore an item or equipment to a stated condition.
- **Preventive maintenance**. These are actions taken in an attempt to retain an item or equipment in a stated condition by providing orderly inspection, detection, and prevention of incipient failure.
- **Failure**. This is the inability of a piece of equipment or item to function within previously stated limits.
- **Downtime**. This is that component of time during which the equipment or item is not in condition to carry out its stated mission.
- **Maintainability function**. This is a plot of the probability of repair within a time stated on the vertical axis vs. maintenance time on the horizontal axis. It is extremely useful to predict the probability that repair will be completed in a specified time.
- **Maintenance plan**. This is a document that outlines the management and technical approach to be employed to maintain an equipment or item.
- **Maintainability parameters**. This is a class of factors or human, environmental, and design features that influence the carrying out of maintenance on product or equipment.
- **Maintainability demonstration**. This is the joint manufacturer and customer effort to determine whether stated maintainability goals have been satisfied.

11.3 MEDICAL EQUIPMENT CLASSIFICATION AND INDICES FOR REPAIR AND MAINTENANCE

Recently the Association for the Advancement of Medical Instrumentation (AAMI) conducted a pilot study to help medical technology managers reduce repair and maintenance costs and enhance the effectiveness of repair/maintenance services.[12] This was achieved by developing common, standardized cost and quality metrics or indices so that repair and maintenance services could be compared among organizations involved with medical equipment. The study focused on three indices and surveyed eight organizations (University of California Davis Medical Center, Sacramento; University Medical Center, Tuscon, AZ; Shands Hospital at the University of Florida, Gainesville; Thomas Jefferson University, Philadelphia; Medical College of Ohio, Toledo; Baylor University Hospital, Dallas; Loma Linda University, Loma Linda; and University of Missouri, Columbia.) and classified equipment into eight categories, as shown in Figure 11.1: patient diagnostic, imaging and radiation therapy, laboratory apparatus, life support and therapeutic, patient environmental and transport, and miscellaneous medical equipment.[12]

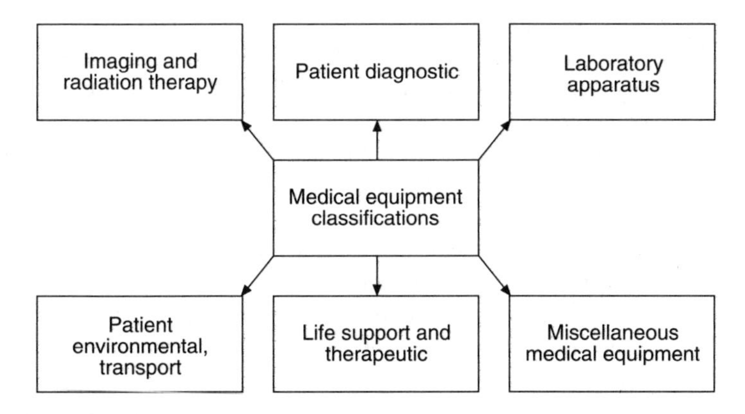

FIGURE 11.1 Medical equipment classification for the purpose of repair and maintenance.

The patient diagnostic classification includes devices connected to the patient and used to collect, record, and analyze information concerning patients. Some examples of such devices are spirometers, physiologic monitors, and endoscopes. The imaging and radiation therapy classification includes devices used to image patient anatomy and radiation therapy equipment. Some examples of the equipment belonging to this category are X-ray machines, ultrasound devices, and linear accelerators.

The laboratory apparatus category includes devices used in the preparation, storage, and analysis of *in vitro* patient specimens. Three examples of such devices are lab analyzers, lab refrigeration equipment, and centrifuges. The life support and therapeutic category of equipment apply energy to the patient. Such equipment includes ventilators, lasers, anesthesia machines, powered surgical instruments, and electrosurgical units.

The patient environmental and transport classification of equipment includes patient beds and other items whose goal is the transport of patients or the improvement of the patient environment. Examples of such equipment are wheelchairs, examination lights, gurneys, and patient-room furniture. The miscellaneous medical equipment classification includes items that are not included in the other five categories, for example, sterilizers.

11.3.1 Repair and Maintenance Indices

The AAMI study focused on three metrics or indices: one cost and two quality.[12]

Cost Index

This is expressed by

$$CR = \frac{SC}{AC}, \tag{11.1}$$

where

- CR is the cost ratio.
- SC is the service cost or the total of all labor, parts, and material costs for scheduled and unscheduled service, including in-house, vendor, prepaid contracts, and maintenance insurance.
- AC is the acquisition cost or the cost at the time of purchase of equipment.

Four advantages cited for using the index are easy to compare across equipment types, can be used with incomplete data, are usually used by vendors to price service contracts, and take into account all service costs. In contrast, index disadvantages include requirement of a standard definition for the pricing of in-house service, no compensation for the age of the equipment, and no mechanism to adjust wage rates by geographic region.

In the organizations that responded to the survey, the service and acquisition costs varied from $454,499 to $3,801,202 and from $19 million to approximately $70 million, respectively. Consequently, the value of the cost ratio (CR) ranged from 2.1% to 5.5%. However, the average values of service cost, acquisition cost, and CR were $42.193 million, $1.653 million, and 3.9%, respectively.

Table 11.1 presents a range of values of CR and its average value for various classifications of medical equipment surveyed.

Quality Index I

This index provides repair requests completed per device; thus it is analogous to equipment repair rate. More specifically, it measures how frequently the customer has to request service per device and is expressed by

$$RR = \frac{NRRC}{\theta}, \tag{11.2}$$

TABLE 11.1
A Range of Values of CR and Its Average Value for Various Classifications of Medical Equipment

No.	Medical Equipment Classification	CR Values Range (%)	CR Values Average (%)
1	Laboratory apparatus	1.9–8.6	5.1
2	Imaging and radiation therapy	1.0–6.7	5.6
3	Patient diagnostic	1.7–3.8	2.6
4	Life support and therapeutic	2.3–5.3	3.5
5	Patient environmental and transport	1.4–8.5	4.4
6	Miscellaneous medical equipment	1.6–3.9	2.6

where

RR is the number of repair requests completed per device.
$NRRC$ is the total number of repair requests.
θ is the number of devices or pieces of equipment.

In the survey, the value of RR ranged from 0.3 to 2.0, with a mean of 0.8.

Quality Index II

This index gives average turnaround time per repair; thus it measures how much time elapses from a customer request until the failed device or equipment is repaired and put back in service. The index is defined as follows:

$$ATAT = \frac{TTAT}{\alpha}, \tag{11.3}$$

where

$ATAT$ is the average turnaround time per repair.
$TTAT$ is the total turnaround time.
α is the total number of work orders or repairs.

Only five hospitals provided data for turnaround times, and the average value of ATAT was 79.5 hours. However, among these hospitals, the turnaround time per repair ranged from 35.4 to 135 hours.

11.4 COMPUTERIZED MAINTENANCE MANAGEMENT SYSTEM FOR MEDICAL EQUIPMENT AND DEVICES AND ITS SELECTION

Clinical engineering departments in hospitals use Computerized Maintenance Management Systems (CMMS) to collect, store, analyze, and report data on the repair and maintenance performed on medical equipment and devices. In turn, these data are used for various purposes, including work order control, equipment management, cost control, quality improvement activities, and reliability and maintainability studies. Over the years, CMMS for clinical engineering have become so complex and sophisticated that many clinical engineering departments find them too costly and time-consuming to develop, update, and maintain internally. Nowadays, there are many commercially available CMMS that can be used by the Clinical Engineering departments. Table 11.2 presents names and addresses of selective companies that have developed such systems.[13]

TABLE 11.2
Companies that Market Computerized Maintenance Management Systems

No.	Company Name and Address
1	Maintenance Automation Corporation 3107 W. Hallandale Beach Blvd. Hallandale, FL 33009, USA
2	Butterfield Systems 11225 W. Bernardo Court San Diego, CA 92127, USA
3	American Services Resources 24331 Muirlands Blvd., Bldg. 4, Suite 125 Lake Forest, CA 92630, USA
4	CHAMPS Software 1255 N. Vantage Pt. Dr. Crystal River, FL 34429, USA
5	Facility Management Systems 8145 River Drive Morton Grove, IL 60053, USA
6	CK Systems Inc. 772 Airport Blvd. Ann Arbor, MI 48108, USA

A similar process to the prepurchase evaluation of a medical equipment can be used in selecting a commercial CMMS. The major elements of this process are as follows:[13]

- Define the problem scope.
- Evaluate the current system being used.
- Perform a preliminary study of commercially available systems.
- Perform a comprehensive study of the chosen CMMS.
- Discuss potential performance issues related to the selected CMMS.
- Examine CMMS support issues.
- Examine CMMS cost.

Defining the problem scope is the first step of the CMMS selection process. An example of the problem scope is: the clinical engineering department needs to procure a CMMS to collect, store, analyze, and report device repair data more efficiently. The second step of the CMMS selection process is evaluating the current system being used. This could be accomplished by answering questions such as:[13]

- What types of data currently exist?
- What are the sources of the current data?
- How does the current data reach the clinical engineering department?
- How is the current data being utilized?
- What elements does the current system not possess?
- What are the projected needs?
- Are any unique or special requirements currently in action?
- Is there any need to specify other special or unique requirements to meet potential needs?

In the third step — performing a preliminary study of commercially available systems — the commercially available CMMS are evaluated simply by reading marketing literature and understanding the general focus and philosophy of each system. The results of this evaluation can be used to develop a preliminary budget and to determine which system to focus on for a comprehensive examination.

In the fourth step — performing a comprehensive study of the chosen CMMS — the commercially available CMMS are evaluated carefully with respect to identified requirements. In this case, similar approaches used in prepurchase evaluation of medical devices/equipment can be employed. Nonetheless, one must remember that the flexibility of CMMS is important, and it must be considered with care.

The other major element of the selection process is to discuss potential performance issues related to the selected CMMS. As the demonstration software programs are single-user and restricted in size, they are ineffective in providing a good indication of large database performance or multi-user performance. Thus, after identifying a small group of candidates that appear to meet specified requirements, vendors should be asked to provide names of customers of similar size and scope that have used the CMMS under consideration. The results of the discussions with these reference customers should be used to determine whether the performance of systems will effectively satisfy the in-house needs.

The next important element of the selection process is to examine CMMS support issues. This basically involves examining the support services provided by the vendor. Nonetheless, vendor support may be categorized into two groups: initial start-up support and continuing support. The elements of the initial start-up support include initial assistance for software and hardware installation, user training, initial data entry of codes and inventory data, initial system documentation, and downloading of pre-existing data. Similarly, the elements of the continuing support are periodic software enhancements and "bug" fixes, telephone support (i.e., 800 number), electronic bulletin board, etc.

Lastly, "examine CMMS cost" is another important element of the selection process. This involves developing preliminary budget and cost estimates. Such estimates can be used to perform life cycle cost analysis of the selected group of "finalists." Some examples of the costs involved are computer hardware costs, including a file server and client workstations; initial CMMS software cost; cost of initial network operating system license; annual software support cost; cost of initial training; and cost of peripherals, bar code readers, and supplies (e.g., disks).

11.5 VENTILATOR MAINTENANCE AND FIELD PERFORMANCE

The material presented in this section is the result of a U.K. study of 11 lung ventilators used in the Intensive Care Unit of the Wakefield General Hospital.[14] At the time of the study, the hospital had the planned preventive maintenance (PPM) system under which all the ventilators underwent maintenance every six weeks, half yearly, and at yearly intervals. The history record sheets of the ventilators contained information such as the following:[14]

- Type of service or repair
- Date of service or repair
- Service/repair time
- Description of repair parts used
- Part and labor costs

In addition, all ventilators were installed with running/frequency meters to determine their utilization. The study concluded factors such as these:

- A total of 131 faults occurred during the period of 14,066 calendar operating days for all ventilators.
- On average, 16 hours were spent annually (per ventilator) for maintenance, out of which only 12.3% accounted for breakdown maintenance.
- Statistically one in every 240 ventilators is not available for service either due to breakdowns or maintenance.
- The potential risk of death due to faulty ventilators is one in every 75,000 patients.
- The efficiency of the PPM is around 65%, i.e., PPM helped to remove 65% of the potential faults that otherwise would have resulted in a failure.

11.6 MODELS FOR MEDICAL EQUIPMENT MAINTENANCE

Over the years, many mathematical models concerning engineering equipment maintenance have been developed. Many of these models can be applied in performing medical equipment maintenance. This section presents three such models.[15-18]

11.6.1 Spare Part Prediction Model

This is a useful mathematical model to predict the number of spares required. Thus, the number of spares required for each item in use is expressed by

$$NS = \lambda T + z[\lambda T]^{1/2} , \tag{11.4}$$

where

- NS is the number of spares.
- λ is the constant failure rate of the item under consideration.
- T is the mission time.
- z is associated with the cumulative normal distribution function; its value is dependent upon a specified confidence level for "no stock out." Thus, for a given confidence level, the value of z is obtained from the standardized cumulative normal distribution function table given in various mathematical and other books. Nonetheless, the standardized cumulative normal distribution function is defined by

$$P(z) = \frac{1}{\sqrt{2\pi}} \int_{-\infty}^{z} e^{-y^2/2} dy . \tag{11.5}$$

Example 11.1

After performing failure data analysis, it was concluded that the times to failure of ventilators are exponentially distributed with the mean of 5000 hours. Calculate the number of spare ventilators required if the mission time is 4000 hours and the confidence level for "no stock out" of the ventilators is 0.8413.

Thus, the constant failure rate of ventilators is

$$\lambda = \frac{1}{5000} = 0.0002 \text{ failures/hour} .$$

For the confidence level of 0.8413 by using the table of the standardized cumulative normal distribution function, we get

$$z = 1.$$

Using the above calculated values and the given mission time of 4000 hours in Equation (11.4) yields

$$NS = (0.0002)(4000) + (1)[0.0002(4000)]^{1/2}$$

$$\simeq 2 \text{ ventilators} .$$

It means two spare ventilators are needed.

11.6.2 Optimum Inspection Estimation Model

This model is developed to estimate the optimum number of inspection per equipment per unit time. The determination of the optimum number of inspections is important because the inspection is often disruptive. But on the other hand, it help reduce equipment downtime because of fewer unexpected breakdowns. The model minimizes the total equipment downtime to obtain the optimum number of inspections.

The total equipment downtime is expressed by

$$TDT = k\,T_{di} + c\,T_{db} k^{-1}, \tag{11.6}$$

where

TDT is the total equipment downtime per unit time.
T_{di} is the downtime per inspection for a piece of equipment.
T_{db} is the downtime per breakdown for the piece of equipment under consideration.
c is a constant for specific facility under consideration.
k is the number of inspections per piece of equipment per unit time.

By differentiating Equation (11.6) with respect to k and then equating it to zero, we get

$$T_{di} - \frac{CT_{db}}{k^2} = 0. \tag{11.7}$$

Rearranging Equation (11.7) yields

$$k^* = \left[CT_{db}/T_{di}\right]^{1/2}, \tag{11.8}$$

where

k^* is the optimum number of inspections per piece of equipment per unit time.

Example 11.2

Assume that the following data are associated with a X-ray machine:

- T_{db} = 0.20 months
- T_{di} = 0.04 months
- c = 2

Determine the optimum number of inspections to be performed per month so that the total downtime of the X-ray machine is at minimum.

Inserting the given data into Equation (11.8) yields

$$k^* = [(2)(0.20)/0.04]^{1/2}$$
$$= 3.16 \text{ inspections per month.}$$

Thus, the optimum number of inspections to be performed per month is 3.16.

11.6.3 Optimum Time Between Replacements Estimation Model

This is a useful model to determine the optimum time between replacements. The model also minimizes the average annual total cost with respect to the life of the equipment/item. The item/equipment average annual total cost is made up of three components: average investment cost, average operating cost, and average maintenance cost. Thus, we write

$$TC = OC_1 + MC_1 + \frac{IC}{t} + \frac{(t-1)}{2}(j_{OC} + j_{mc}), \tag{11.9}$$

where

- TC is the average annual total cost of item/equipment.
- t is the equipment/item life in years.
- IC is the investment cost.
- OC_1 is the item/equipment operational cost for the first year.
- MC_1 is the item/equipment maintenance cost for the first year.
- j_{OC} is the amount by which operational cost increases annually.
- j_{mc} is the amount by which maintenance cost increases annually.

By differentiating Equation (11.9) with respect to t and then equating it to zero, we get

$$t^* = \left[\frac{2IC}{j_{OC} + j_{mc}}\right]^{1/2}, \tag{11.10}$$

where

- t^* is the optimum-replacement interval.

Example 11.3

The following data were obtained on a certain piece of medical equipment:

$$IC = \$200,000$$
$$j_{OC} = \$5000$$
$$j_{mc} = \$2000$$

Determine the optimum replacement interval for the equipment.

Substituting the specified values into Equation (11.10) yields

$$t^* = \left[\frac{2(200,000)}{(5000)+(2000)}\right]^{1/2}$$

$$= 7.6 \text{ years}.$$

Thus, the optimum replacement interval for the medical equipment is 7.6 years.

11.7 MAINTAINABILITY

The application of maintainability principles in designing engineering equipment has helped to produce effectively maintainable products. Its careful application during the design of medical equipment/devices can also help produce effective medical equipment with respect to maintainability. Reasons for applying maintainability engineering principles could be to reduce projected maintenance time and cost through design modifications, to determine the downtime due to maintenance, to determine the amount of labor hours and related resources required to perform the projected maintenance, etc.[19]

11.7.1 MAINTAINABILITY DESIGN FACTORS

The most frequently addressed factors include the following:[2]

- Accessibility
- Controls
- Displays
- Test points
- Labelling and coding
- Manuals, checklists, charts, and aids
- Test equipment
- Connectors
- Cases, covers, and doors
- Mounting and fasteners
- Handles

There are many other factors, for example, modularization, interchangeability, standardization, ease of removal and replacement, illumination, installation, adjustments and calibrations, lubrication, weight, skill requirements, work environment, and training requirements.[2] Four of these factors are described below.

Accessibility

This may simply be described as the relative comfort with which a part can be reached for actions such as service, repair, or replacement. Some of the resulting factors of poor accessibility design are increased repair time, human error, injuries, accidental equipment damage, and so on. Many factors affect accessibility, including the type of maintenance task to be performed, access usage frequency, distance to be reached, work clearances needed to perform required functions satisfactorily, types of tools required for maintenance actions, and specified time requirements for maintenance activities.[2,20] Equipment designers should consider the following guidelines with respect to ease of maintenance:

- Design only one access to remove any replaceable unit.
- Locate each item independently so that it is easily reached.
- Locate each item so that access to it is not blocked by structural items.
- Avoid locating items or parts beneath floor boards, the operator's seat, pipes, hoses, and structural members.
- Emphasize the design in plug-in modules.

Modularization

This may simply be described as the division of an equipment or system into distinct physical and functional units to allow removal and replacement. Modularization should be used whenever possible and, in particular where its application would lead to reduction in personnel training. However, factors such as cost, practicality, and function determine the degree of modularization. Advantages of modularization include easily maintainable product or equipment, reduction in maintenance time and cost, relatively low skill levels and fewer tools needed, and simplified new design, thus shorter design time.

Some important guidelines associated with the modularization design are as follows:[20-21]

- Divide the equipment/system under consideration into many modules.
- Strive to make modules as uniform in size and shape as feasible.
- Aim to make each module capable of being examined independently.
- Strive to design the entire equipment so that an individual can replace any failed part.
- Ensure that each module is small and light enough so that an individual can handle and carry it in an effective manner.
- Design the module units so that their operational testing is easy and straightforward when they are separated from the equipment.

Standardization

This is an important design feature that imposes restrictions on the variety of parts to be used in a piece of equipment or item. Standardization should be the main goal of design because the use of nonstandard parts may lead to lower reliability and increased maintenance, but it must not be permitted to interfere with improvements in design or advances in technology. Basic goals of standardization include maximizing the use of common components in different items or equipment; maximizing the use of interchangeable parts and components; reducing the number of different types of parts, assemblies, and other items; reducing the number of models and makes of equipment in use; and reducing storage problems.

Advantages of standardization are as follows:

- Reduction in manufacturing cost and design time
- Improvement in item reliability
- Less need for special or close tolerance parts
- Reduction in maintenance time and cost
- Less chance of accidents due to incorrect or unclear procedures
- Reduction in the danger of misuse of parts
- Reduction in acquisition, stocking, and training related problems

Interchangeability

Interchangeability is closely related to standardization, and it simply means that a given item can be replaced by any like item. The basic factors to be considered in determining interchangeability requirements are field use conditions and economy of manufacture and inspection. Three basic principles of interchangeability are as follows:[2,18,20]

- Liberal allowance for tolerances
- Every part or component is totally interchangeable with every other like part in equipment or items requiring frequent replacement and servicing of component parts because of wear or damage
- Strict interchangeability may not be cost-effective in items that are normally expected to function without part replacement

11.8 MAINTAINABILITY MEASURES

In maintainability analysis of engineering systems/equipment, various types of maintainability measures are used. Such measures can be applied equally to perform maintainability analysis of medical equipment or devices. Three of these measures are described below.[2,18,19,22,23]

11.8.1 Mean Time to Repair (MTTR)

This is probably the most widely used maintainability measure. It measures the elapsed time required to conduct a specified maintenance activity. MTTR is expressed by

$$MTTR = \left(\sum_{i=1}^{k} T_{ri} \lambda_i \right) \Big/ \sum_{i=1}^{k} \lambda_i , \qquad (11.11)$$

where

k is the total number of units or items.
T_{ri} is the repair time required to repair item or unit i; for i = 1, 2, 3, ..., k.
λ_i is the failure rate of item or unit i; for i = 1, 2, 3, ..., k.

Although time to repair may be represented by various statistical distributions, the exponential distribution is usually assumed for electronic equipment having an effective built-in test capability along with a rapid remove-and-replace maintenance concept. However, for electronic equipment without a built-in test capability, normally the log normal distribution is assumed for times to repair.

Example 11.4

A medical equipment is made up of five subsystems whose failure rates and corrective maintenance times are given in Table 11.3. Calculate MTTR of the medical equipment.

TABLE 11.3
Medical Equipment Subsystem Failure Rates and Corrective Maintenance Times

Subsystem No.	Failure Rate (per hour)	Corrective Maintenance Time (hours)
1	0.0005	1
2	0.0006	1.5
3	0.0007	2
4	0.0008	2.5
5	0.0009	3

Using the given data in Equation (11.11) yields

$$MTTR = \frac{(1)(0.0005)+(1.5)(0.0006)+(2)(0.0007)+(2.5)(0.0008)+(3)(0.0009)}{(0.0005)+(0.0006)+(0.0007)+(0.0008)+(0.0009)}$$

$$= 2.1429 \text{ hours}.$$

Thus, the medical equipment mean time to repair is 2.1429 hours.

11.8.2 Mean Preventive Maintenance Time

This is another important measure of maintainability, as preventive maintenance of some sort is usually performed on engineering equipment. For example, activities such as inspections, tunning, and calibrations are performed to keep equipment at a specified performance level. The mean preventive maintenance time is defined by

$$MPMT = \frac{\sum_{i=1}^{n} (ETPMT_i)(FPMT_i)}{\sum_{i=1}^{n} FPMT_i}, \tag{11.12}$$

where

- $MPMT$ is the mean preventive maintenance time.
- n is the total number of preventive maintenance tasks.
- $FPMT_i$ is the frequency of preventive maintenance task i; for i = 1, 2, 3, ..., n.
- $ETPMT_i$ is the elapsed time for preventive maintenance task i; for i = 1, 2, 3, ..., n.

11.8.3 Maintainability Function

This is a useful measure to predict the probability that the repair will be accomplished in a time, t, when it starts on an item at time t = 0. The maintainability function, m (t), is defined by

$$m(t) = \int_0^t f(t)\,dt, \tag{11.13}$$

where

- t is time.
- $f(t)$ is the probability density function of the repair time.

The maintainability functions for various probability distributions may be obtained by using their probability density functions in relationship (11.13).

Example 11.5

Assume that the time to repair an X-ray machine is described by the following probability density function:

$$f(t) = \frac{1}{MTTR} \exp\left(-\frac{t}{MTTR}\right), \tag{11.14}$$

where

- t is time.
- $MTTR$ is the mean time to repair of the X-ray machine.

If MTTR = 2 hours, determine the probability that a repair will be accomplished in four hours.

Using Equation (11.14) in Equation (11.13) yields

$$m(t) = 1 - \exp\left[-\frac{t}{MTTR}\right]. \tag{11.15}$$

By substituting the given data into Equation (11.15), we get

$$m(4) = 1 - \exp\left[-\frac{4}{2}\right]$$

$$= 0.8647.$$

It means, there is approximately an 87% chance that the repair will be completed in four hours.

11.9 PROBLEMS

1. Define the terms "maintenance" and "maintainability."
2. Discuss the following classifications of the medical equipment:
   - Imaging and radiation therapy
   - Patient diagnostic
   - Life support and therapeutic
3. What is the "computerized maintenance management system" with respect to medical devices/equipment?
4. List the major elements of the computerized maintenance management system selection process used in the medical field.
5. What is a ventilator?
6. The times to failure of an X-ray machine part are exponentially distributed with a mean of 3000 hours. Determine the number of spare X-ray machine parts required if the value of the mission is 2000 hours and the confidence level for "no stock out" of the parts is 0.9772.
7. What are the most frequently addressed maintainability design factors?
8. What are the benefits of modularization?
9. Discuss the following two terms:
   - Standardization
   - Accessibility
10. Assume that the times to repair a medical equipment follow log normal probability density function. Obtain an expression for its maintainability function.

REFERENCES

1. Shooman, M.L., *Probabilistic Reliability: An Engineering Approach*, McGraw-Hill Book Company, New York, 1968.

2. AMCP-133, *Engineering Design Handbook: Maintainability Engineering Theory and Practice*, Department of the Army, Washington, DC, 1976.
3. McKenna, T. and Oliverson, R., *Glossary of Reliability and Maintenance*, Gulf Publishing Company, Houston, TX, 1997.
4. Norman, J.C. and Goodman, L., Acquaintance with and Maintenance of Biomedical Instrumentation, *J. Assoc. Advan. Med. Inst.*, Vol. 1, September 1966, pp. 8–10.
5. Waits, W., Planned Maintenance, *Med. Res. Eng.*, Vol. 7, No. 12, 1968, pp. 15–18.
6. Dhillon, B.S., *Reliability Engineering Applications: Bibliography on Important Application Areas*, Beta Publishers, Inc., Gloucester, Ontario, Canada, 1992.
7. Dhillon, B.S. and McCrea, J.L., Bibliography of Literature on Medical Equipment Reliability, *Microelectronics and Reliability*, Vol. 20, 1980, pp. 737–742.
8. Dhillon, B.S., *Engineering Maintainability*, Gulf Publishing Company, Houston, TX, 1999.
9. Omdahl, T.P., Ed., *Reliability, Availability, and Maintainability (RAM) Dictionary*, ASQC Quality Press, Milwaukee, WI, 1988.
10. MIL-STD-721, Definitions of Effectiveness Terms for Reliability, Maintainability, Human Factors, and Safety, Department of Defense, Washington, DC.
11. Von Alven, W.H., Ed., *Reliability Engineering*, Prentice-Hall, Inc., Englewood Cliffs, NJ, 1964.
12. Cohen, T., Validating Medical Equipment Repair and Maintenance Metrics: A Progress Report, *Biomedical Instrumentation and Technology*, Jan./Feb., 1997, pp. 23–32.
13. Cohen, T., Computerized Maintenance Management Systems: How to Match Your Department's Needs With Commercially Available Products, *J. Clin. Eng.*, Vol. 20, No. 6, 1995, pp. 457–461.
14. Keller, A.Z., Kamath, A.R.R., and Peacock, S.T., A Proposed Methodology for Assessment of Reliability, Maintainability and Availability of Medical Equipment, *Reliability Engineering*, Vol. 9, 1984, pp. 153–174.
15. Ebel, G. and Lang, A., Reliability Approach to the Spare Parts Problem, Proceedings of Ninth National Symposium on Reliability and Quality Control, 1963, pp. 85–92.
16. Dhillon, B.S., *Mechanical Reliability: Theory, Models, and Applications*, American Institute of Aeronautics and Astronautics, Inc., Washington, DC, 1988.
17. Wild, R., *Essentials of Production and Operations Management*, Holt, Rinehart and Winston, London, 1985.
18. Dhillon, B.S., *Engineering Maintainability*, Gulf Publishing Company, Houston, TX, 1999.
19. Grant-Ireson, W., Coombs, C.F., and Moss, R.Y., *Handbook of Reliability Engineering and Management*, McGraw-Hill, New York, 1988.
20. AMCP 706-134, *Engineering Design Handbook: Maintainability Guide for Design*, Department of Defense, Washington, DC, 1972.
21. Altman, J.W., et al., *Guide to Design of Mechanical Equipment for Maintainability*, Report No. ASD-TR-61-381. U.S. Air Force Systems Command, Wright-Patterson Air Force Base, Ohio, 1961.
22. Blanchard, B.S., *Logistics Engineering and Management*, Prentice-Hall Inc., Englewood Cliffs, NJ, 1981.
23. Blanchard, B.S., Verma, D., and Peterson, E.L., *Maintainability*, John Wiley & Sons, Inc., New York, 1995.

12 Reliability-Related Standards, Failure Data Sources, and Failure Data and Analysis for Medical Devices

CONTENTS

12.1 INTRODUCTION

As we make strides in technological developments, the demand for medical devices is increasing at an alarming rate throughout the world. The history of the use of medical devices may be traced back to ancient times. For example, dental devices were used by the ancient Egyptians and Etruscans.[1-2] However, the arrival of Franz Anton Mesmer in Paris in 1778 may be considered the beginning of exaggerated medical claims with mechanical and electrical devices. For example, he claimed that "animal magnetism," the primary "agent of nature," was the source of all health.[2-6]

Nowadays, due to various reasons, including to counter false claims, many standards and documents are used in the development of medical devices. For example, in 1987 the American National Standards Institute's (ANSI) annual report of medical device standards board activities identified more than 700 voluntary medical standards completed or under development.[7] These standards address issues directly or indirectly relating to medical device reliability and safety: safety testing, electrical safety, leakage current, performance, manufacturing requirements, etc.

Just as in the case of any other engineering products, the failure data are the backbone of medical device reliability studies. Such data provide invaluable information to professionals such as reliability and design engineers involved with medical devices, because the information is the final proof of reliability-related efforts expended during device design and manufacture.

This chapter presents different aspects of reliability-related standards and failure data for medical devices.

12.2 WHAT IS A STANDARD?

There are many definitions to the word "standard," and usually, for example, the first definition is "anything taken as a basis of comparison."[7] This definition includes the most general manifestations of standards as visualized by the product manufacturers: approved and accepted practices, specifications, guidelines, reference method and materials, etc. Standards are available on many diverse areas, and they can fit into categories such as:[7]

- **Recommended practice**. This outlines an accepted procedure for performing something.
- **Testing method**. This is an accepted approach to measuring something.
- **Specification**. This sets limits on the characteristics of a product or material. It may be divided into two categories: design and performance.

Design specification is important because it promotes uniformity and compatibility, or put simply, it specifies the ingredients of the product. On the other hand, the performance specification sets limits on the product performance.

12.3 ORGANIZATIONS INVOLVED WITH DEVELOPING MEDICAL DEVICE STANDARDS AND OTHER DOCUMENTS

Many organizations are involved with the development of medical device-related standards and other documents. This section describes briefly some of these organizations and includes a list of nondescribed organizations.[8-12]

12.3.1 ASSOCIATION FOR THE ADVANCEMENT OF MEDICAL INSTRUMENTATION (AAMI)

The Association for the Advancement of Medical Instrumentation (AAMI) has a membership of 5000 technology developers, users, and managers with common desire to share their knowledge for the improvement of medical technology.[8] It serves its membership and the field well by providing items such as the following:

- Standards on equipment safety, performance, development, and use
- Educational programs
- Publications such as "Medical Instrumentation" and "Biomedical Technology Today"
- Personal interactions among its members through its annual meeting

In 1972, the AAMI organized a conference on medical device standards that subsequently played an instrumental role in developing basic concepts of the AAMI standards program. In fact, ever since the conference, the AAMI has published several standards, recommended practices, and technical information reports. Table 12.1 presents some of these documents directly or indirectly related to medical device reliability/safety.[8]

12.3.2 FOOD AND DRUG ADMINISTRATION (FDA)

This is an important regulatory body in the U.S. with respect to medical devices. Although the Food and Drug Administration (FDA) enacted its first regulations concerning public health in 1906, prior to 1976 it had only limited authority over medical devices. After the passage of Medical Device Amendments to the Federal Food, Drug and Cosmetic Act (FFD and CA) of 1938 by the U.S. Congress, the FDA's authority increased substantially over medical devices.

In 1978, the FDA introduced Good Manufacturing Practices (GMPs) regulation, which basically represented a good quality assurance program to control the manufacture, package, distribution, storage, and installation of medical devices in the U.S. The other important regulations to medical devices were as follows:[12]

- **1984**: Medical device reporting (MDR)
- **1988**: Device reconditioner/rebuilder (DRR)
- **1992**: Safe Medical Devices Act (SMDA)

TABLE 12.1
AAMI Selective Documents Directly or Indirectly Related to Medical Device Reliability

No.	Document Title	Document Reference No.
1	Human factors engineering guidelines and preferred practices for the design of medical devices (AAMI recommended practice)	AAMI HE-1988
2	Safe current limits for electro medical apparatus (American National Standard)	ANSI/AAMI ES1-1985
3	Establishing and administering medical instrumentation maintenance programs, guidelines for (AAMI recommended practice)	AAMI MIM-3/84
4	Testing and reporting performance results of ventricular arrhythmia detection algorithms (AAMI recommended practice)	AAMI ECAR-1987
5	Diagnostic electro cardiographic devices (American National Standard)	ANSI/AAMI EC11-1982
6	Automatic external Defibrillator design, testing and reporting performance results (AAMI technical information report)	AAMI TIR No. 2-1987
7	Process control guidelines for gamma radiation sterilization of medical devices (AAMI recommended practice)	AAMI RS-3/84
8	Performance evaluation of ethylene oxide sterilizers - EO test packs, good hospital practice (AAMI recommended practice)	AAMI EOTP-2/85

Although the FFD and CA and FDA have most of the authority over medical devices, time to time medical devices do fall within the purview of other federal and state bodies. Some of these bodies are as follows:[11]

- **Federal Communications Commission (FCC).** The FCC regulates medical devices that emit electromagnetic energy on frequencies within the framework of the radio frequency spectrum. Examples of such devices are diathermy and ultrasonic equipment.
- **Occupational Safety and Health Administration (OSHA).** OSHA has the authority to examine medical device facilities and impose appropriate fines when the violation of the law occurs.
- **Consumer Product Safety Commission (CPSC).** The CPSC enforces the Federal Hazardous Substances Act (FHSA) by establishing labelling and warning requirements for products that are toxic, flammable, sensitizers, corrosive, or irritants. Although certain medical devices could be subject to the FHSA, in practice, the CPSC's authority over medical devices is rather limited.

- **Federal Trade Commission (FTC).** The FTC regulates the advertising of nonrestricted medical devices and has developed standards for "comparative claims", which compare the characteristics of competing products or items.
- **Environmental Protection Agency (EPA).** The EPA also has jurisdiction over medical devices because substances generated from the production of these items may be released into the air, water, and soil.

12.3.3 International Electrotechnical Commission (IEC)

The International Electrotechnical Commission (IEC) came into being in 1906 for the purpose of developing international standards within the framework of electrical and electronic areas. Over the years, it has developed many standards. Some of its standards that are directly or indirectly related to the medical device reliability/safety are presented in Table 12.2.[9]

TABLE 12.2
IEC Standards Directly or Indirectly Related to Medical Device Reliability/Safety

No.	Standard Title	Standard Reference No.
1	Medical electrical equipment Part I: general requirements for safety	IEC 601-1: 1988
2	Medical electrical equipment Part I: general requirements for safety: collateral standard: safety requirements for medical electrical systems	IEC 601-1-1: 1992
3	Reliability testing compliance test plans for success ratio	IEC 1123
4	Lung ventilators for medical use	IEC 601-2-12: 1988
5	External cardiac pacemakers	IEC 601-2-31: 1994
6	Equipment reliability testing	IEC 605
7	Diagnostic X-ray generators	IEC 601-2-7: 1987
8	Diagnostic and therapeutic laser equipment	IEC 601-2-22: 1992
9	Magnetic resonance imaging equipment	IEC 601-2-32: 1994
10	Medical electrical equipment Part I: general requirements for safety: collateral standard: requirements for programmable electronic medical systems	IEC 601-1-4
11	Hospital beds	IEC 601-2-38: 1995
12	Medical electrical equipment Part I: general requirements for safety: collateral standard: electromagnetic compatibility	IEC 601-1-2

12.3.4 American National Standards Institute (ANSI)

Over the years, the American National Standards Institute (ANSI) has produced more than 11,000 standards, and in its 1987 Annual Report on Medical Device Standards Board Activities, it has identified that more than 700 voluntary medical device standards were either completed or under development.[7] The ANSI is the U.S. representative at the IEC and at the International Organization for Standardization.

12.3.5 International Organization for Standardization (ISO)

The International Organization for Standardization (ISO) was established in 1947 for standardization in general; however, the development of electrical and electronic standards is still the sole responsibility of the IEC. Members of the ISO are the national standards organizations of close to 90 countries, and ISO secretariats support 162 technical committees and approximately 600 subcommittees in 32 countries.[10] Some of the ISO standards directly or indirectly concerning medical device reliability/safety are listed in Table 12.3.[10]

TABLE 12.3
Selective ISO Standards Directly or Indirectly Concerning Medical Device Reliability/Safety

No.	Standard Title/Description	Standard Reference No.
1	ISO 9000 series of standards (This is a series of internationally recognized standards specifying a quality management system.)	ISO 9000, ISO 9001, ISO 9002, ISO 9003, and ISO 9004
2	Quality systems — medical devices — particular requirements for the application of ISO 9001	ISO 13485
3	Quality systems — medical devices — particular requirements for the application of ISO 9002	ISO 13488
4	Quality systems — medical devices — Guidance on the application of ISO 13485 and ISO 13488	ISO/DIS 14969

12.3.6 Department of Defense

Over the years, the U.S. Department of Defense has produced effective standards and other documents for use in military procurement. In fact, from time to time such material is also used for nonmilitary procurements. Furthermore, many of these documents can equally be used in performing reliability studies concerning medical devices. Some of these documents are presented in Table 12.4.

12.3.7 Institute of Electrical and Electronics Engineers (IEEE)

The Institute of Electrical and Electronics Engineers (IEEE) is the largest professional body in the world. It has produced various types of documents directly or indirectly related to medical device reliability/safety. IEEE was formed in 1884, and its recent emphasis has been on the developments of documents/standards in the area of software development and quality assurance. Some of its software standards have been accredited by the ANSI and have also been referenced by the FDA in the development of guidelines on software for medical purposes.[12]

TABLE 12.4
Selective Documents Produced by the U.S. Department of Defense

No.	Document Title	Document Reference No.
1	Reliability prediction of electronic equipment	MIL-hdbk-217F
2	Definition of terms for reliability and maintainability	MIL-std-721C
3	Reliability modeling and prediction	MIL-std-756B
4	Maintainability prediction	MIL-hdbk-472
5	Reliability program for systems and equipment, development and production	MIL-std-785b
6	Reliability assurance program for electronic parts specifications	MIL-std-790E
7	Electronic reliability design handbook	MIL-hdbk-338
8	Reliability growth management	MIL-hdbk-189
9	Reliability design, qualification and production acceptance tests: exponential distribution	MIL-std-781D
10	Procedures for performing a failure mode, effects, and criticality analysis	MIL-std-1629A
11	Failure reporting, analysis and corrective action system (FRACAS)	MIL-std-2155
12	Sampling procedures and tables for life and reliability testing (based on exponential distribution)	MIL-hdbk-H108
13	Failure classification for reliability testing	MIL-std-2074
14	Reliability test methods, plans, and environments for engineering development, qualification and production	MIL-hdbk-781
15	Failure rate sampling plans and procedures	MIL-std-690C
16	Government/industry data exchange program (GIDEP)	MIL-std-1556B

12.3.8 American Society for Testing and Materials (ASTM)

The American Society for Testing and Materials (ASTM) is the largest source of voluntary consensus standards in the world, and it is a scientific and technical organization whose objective is to develop standards on characteristics and performance of materials/products/services.

Table 12.5 presents some of the standards directly or indirectly related to medical device reliability/safety developed by the ASTM.

12.3.9 Underwriters Laboratory (UL)

The Underwriters Laboratory (UL) is an independent and nonprofit laboratory that examines items such as the following with respect to hazards affecting life and property:[12]

- Devices
- Equipment construction
- Materials
- Methods and systems

TABLE 12.5
Some of the Standards Developed by the ASTM

No.	Standard Title	Standard Reference No.
1	Standard specification for minimum performance and safety requirements for anesthesia breathing systems	ASTM F1208
2	Performance and safety specification for cryosurgical medical instruments	ASTM F0882
3	Specification for alarm signals in medical equipment used in anesthesia and respiratory care	ASTM F1463
4	Test methods for radio-pacity of plastics for medical use	ASTM F0640
5	Guide for computer automation in the clinical laboratory	ASTM F0792

Many health facilities make it mandatory that the medical devices they procure satisfy applicable UL standards. Two UL standards related to medical equipment safety are as follows:

- UL 544: Standard for safety of medical and dental equipment.
- UL 2601: Standard for medical electrical equipment: general requirements for safety.

12.3.10 Miscellaneous organizations

There are many other organizations whose standards/other documents directly or indirectly relate to medical device reliability/safety. Some of these organizations are

- American Society for Quality Control (ASQC)
- National Fire Protection Association (NFPA)
- British Standards Institute (BSI)
- Japanese Standards Association (JSA)
- Institute of Environmental Sciences (IES)
- European Committee for Standardization (CEN)
- European Committee for Electrotechnical Standardization (CENELAC)

12.4 NAMES AND ADDRESSES OF ORGANIZATIONS DIRECTLY OR INDIRECTLY INVOLVED WITH STANDARDS/OTHER DOCUMENTS FOR MEDICAL DEVICES

Many organizations are involved with developing standards and other documents directly or indirectly concerning medical device reliability/safety. Table 12.6 presents the names and addresses of some of these organizations.

12.5 FAILURE DATA AND DATA SOURCES

As in the case of any other engineering product, failure data are the backbone of the medical device reliability studies. The basic objective of failure data collection and analysis is to convert all relevant information accumulated in various sources into an effective form so that it can be efficiently used by professionals with a certain degree of confidence in performing their assigned medical device reliability-related tasks. More specifically, such information may be used in tasks such as design reviews, test planning, quality assurance and inspection planning, and logistic support planning.

During a medical device/equipment life cycle, failure-related data may be collected from many different sources:[13,14] repair facility records, warranty claims, factory acceptance testing, records generated during the development phase, previous experience with similar or identical equipment, inspection records generated by quality control/manufacturing groups, customer failure reporting systems, field demonstration, environmental qualification, and field installation.

12.5.1 Organizations and Sources for Obtaining Medical Device Failure-Related Data

Failure data related to medical devices or to their parts may be obtained from various organizations. Figure 12.1 shows some of those organizations.

Some documents and data banks for obtaining failure data related to medical devices or to their parts are

MIL-HDBK-217, *Reliability Prediction of Electronic Equipment*, Department of Defense, Washington, DC, USA.

This document is often used to predict the failure rate of a given piece of equipment in the industrial sector. It is equally applicable to predict failure rates of medical devices or their parts.

TABLE 12.6
Names and Addresses of Organizations Directly or Indirectly Involved With Developing Standards and Other Documents Concerning Medical Devices

No.	Organization Name	Organization Address
1	Association for the Advancement of Medical Instrumentation (AAMI)	3330 Washington Blvd., Suite 400, Arlington, VA 22201-4598, USA
2	Food and Drug Administration (FDA)	Center for Devices and Radiological Health, 1390 Piccard Drive, Rockville, MD 20850, USA
3	International Electrotechnical Commission (IEC)	Box 131, 3 rue de Varembe, CH-1211 Geneva 20, Switzerland
4	American National Standards Institute (ANSI)	11 West 42nd Street, 13th Floor New York, NY 10036, USA
5	International Organization for Standardization (ISO)	1 rue de Varembe, Case postale 56, CH-1211 Geneva 20, Switzerland
6	Department of Defense (DOD)	Washington, DC, USA
7	Institute of Electrical and Electronic Engineers (IEEE)	455 Hoes Lane, P.O. Box 1331, Piscataway, NJ 08855, USA
8	American Society for Testing Materials (ASTM)	1916 Race Street Philadelphia, PA 19103, USA
9	Underwriters Laboratory (UL)	333 Pfingsten Road, Northbrook, IL 60062, USA
10	American Society for Quality Control (ASQC)	P.O. Box 3005, Milwakee, WI 53201-3005, USA
11	National Fire Protection Association (NFPA)	1 Batterymarch Park, P.O. Box 9101, Quincy, MA 02269-9101, USA
12	British Standards Institute (BSI)	2 Park Street, London W1A 2BS, United Kingdom
13	Japanese Standards Association (JSA)	1-24 Akasaka 4, Minato-ku, Tokyo 107, Japan
14	Institute of Environmental Sciences and Technology (IEST)	940 East North-West Highway, Mount Prospect, IL 60056, USA
15	Canadian Standards Association (CSA)	178 Rexdale Blvd., Rexdale, Ontario M9W 1R3, Canada
16	European Committee for Standardization (CEN)	Rue de Stassart, 36, B-1050 Brussels, Belgium
17	European Committee for Electrotechnical Standardization (CENELAC)	2 rue De Brederode, B-1000 Brussels, Belgium
18	World Health Organization (WHO)	CH-1211 Geneva 27, Switzerland
19	Occupational Safety and Health Administration (OSHA)	200 Constitution Avenue NW, Room N3647, Washington, DC 20210, USA
20	American Hospital Association (AHA)	840 N. Lake Shore Drive, Chicago IL 60611, USA
21	American Medical Association (AMA)	515 N. State Street, Chicago, IL 60610, USA
22	Emergency Care Research Institute (ECRI)	5200 Butler Parkway, Plymouth Meeting, PA 19462, USA

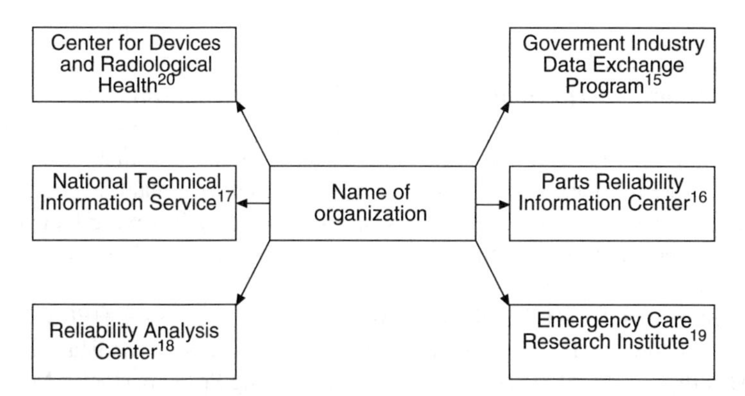

FIGURE 12.1 Some organizations for obtaining failure data related to medical devices or to their parts. *Note*: The numbers following each organization's name refer to the reference numbers at the end of the chapter.

NUREG/CR-1278, *Handbook of Human Reliability Analysis with Emphasis on Nuclear Power Plant Applications*, U.S. Nuclear Regulatory Commission, Washington, DC, USA.

This document provides failure data on various tasks performed by humans, and it could prove to be a useful document in designing medical devices.

Hospital Equipment Control System (HECS™). This computerized system was introduced by ECRI in 1985.[19,21] It:

- Provides objective data on the basis of satisfactory statistical foundation.
- Provides economic, financial, cost-effective, and productivity data within each hospital.
- Serves as the basic building block of a new kind of technology management system.
- Provides efficient support and analysis of clinical engineering operations through detailed work scheduling for inspection, preventive maintenance and repair.

Also HECS™ can provide new information support, including data on the relative reliability of competing brands and models of clinical equipment, repair/replace decision support based on reliable historical data, reliable data on the cost of service, inspection, and preventive maintenance, and better support for technology-related risk management.

Universal Medical Device Registration and Regulatory Management System (UMDRMS). This computerized system was developed by the ECRI,[19,21] and is a universal microcomputer-based stand-alone system. UMDRMS can facilitate many items, including tracking products for recall, safety, and reliability statistics, inventory control, and exchange of information and electronic databases.

Medical Device Reporting System (MDRS).[22] Center for Devices and Radiological Health, FDA, Rockville, MD, USA. This database is managed by the FDA. In 1990, it was expanding by approximately 18,000 reports per year. The database is composed of reports from device manufacturers concerning patient deaths and serious injuries allegedly involving their products, and failures of their manufactured items that may have resulted in death or serious injury.

Miscellaneous ECRI Medical Device Databases.[21] Beside HECS™ and UMDRMS, the ECRI produces many other databases on medical devices: Health Devices Alerts (HDA), Problem Reporting Program (PRP), Medical Device Reporting (MDR), Actual Price Paid Database (APPD), Medical Device Manufacturers Database (MDMD), Health Devices Sourcebook (HDS), and User Experience Network (UEN).

12.5.2 Medical Device Failure-Related Data

This section presents failure data on various items directly or indirectly related to medical devices. Table 12.7 presents data on the preproduction cause of 1143 medical device recalls for 1983–1987.[23]

TABLE 12.7
Breakdown of Preproduction Causes of 1143 Medical Device Recalls for 1983–1987

No.	Cause	No. of Recall Devices	Percentage
1	Device design	386	33.8
2	Process design	318	27.8
3	Component design	270	23.6
4	Device software	80	7.0
5	Label design	49	4.3
6	Package design	38	3.3
7	Process software	2	0.2
		Σ 1143	Σ 100

Table 12.8 presents data on facts and figures concerning medical devices.

Failure rates or failure mode data for selected electronic parts that may find applications in medical devices/equipment are presented in Table 12.9.[12,36]

12.6 FAILURE DATA ANALYSIS

Medical device failure data analysis is equal in importance to failure data collection. Usually the data collected is analyzed to draw intelligent conclusions from it. One

TABLE 12.8
Data on Facts and Figures Concerning Medical Devices

No.	Facts and Figures	Reference(s)
1	The estimate of avoidable fatalities from anesthesia-related incidents is 2000–10,000 per year.	22, 24, 25
2	A U.S. Department of Health, Education, and Welfare special committee estimated that over a ten-year period, 10,000 injuries were associated with medical devices, of which 731 resulted in death. The majority of problems were associated with artificial heart valves (512 deaths and 300 injuries), cardiac pacemakers (89 deaths and 186 injuries), and intra-uterine contraceptive devices (100 deaths and 8000 injuries).	26, 27
3	The committee on hospitals (NFPA) estimated that 1200 deaths per year were due to faulty instrumentation.	28
4	Operator errors account for well over 50% of all technical medical equipment problems.	29
5	An FDA study published in 1990 reported that approximately 44% of the quality-related problems that led to voluntary medical device recall for the period of October 1983 to September 1989 were attributable to deficiencies/errors that could have been eradicated by effective design controls.	30
6	The ECRI reported, after testing a sample of 15,000 hospital products that 4–6% of such products were sufficiently dangerous to warrant immediate correction.	29
7	Human error is responsible for 65–87% of deaths attributable to anaesthesia.	31– 33
8	A study of 212 devices implanted for at least two years indicated their reliability of 0.971 and failure rate of 0.0189 failures per year.	34
9	A study of medication errors made by registered nurses in a hospital indicated that on average, a nurse made one error for every six medications given.	35

important piece of information sought from the collected failure data is the time to failure distribution of the failed items. Over the years, many methods and techniques have been developed that directly or indirectly deal with determining the time to failure distribution and the values of its associated parameters from a given set of data.

This section presents two such methods: one to determine the time to failure distribution, and the other to estimate the distribution parameters.[37-41] Furthermore, both approaches can be used to perform failure data analysis of medical devices.

12.6.1 Bartlett Method

This is a good method to test if a given set of data follow the exponential distribution. Normally, the method is referred to as the Bartlett test, and the test statistic is defined as

TABLE 12.9
Failure Rates or Failure Mode Data for Selected Electronic Parts

No.	Electronic Part	Failure Rate (Failures/ 10^6 hours)	Failure Mode	Occurrence Probability
1	Single fiber optic connectors	0.10	—	—
2	Neon lamps	0.20	—	—
3	Vibrators (MIL-V-95) (60-Cycle)	15.00	—	—
4	Fiber-optic cables (single fiber types only)	0.1 per fiber Km	—	—
5	Microwave ferrite devices: isolators and circulators (≤ 100 Watts and use environments: ground, benign)	0.10		
6	Dummy loads (< 100 Watts and use environments: ground, benign)	0.010	—	—
7	Resister (carbon composition)	—	• Open circuit	0.5
			• Resistance value drift	0.5
8	Capacitor (ceramic)	—	• Short circuit	0.99
			• Open circuit	0.01
9	Micro switch	—	• Open circuit	0.30
			• High resistance	0.60
			• No function	0.1
10	Push-button switch	—	• Short	0.07
			• Open	0.60
			• Sticking	0.33
11	Relay	—	• Contact failure	0.90
			• Coil Failure	0.10
12	Keyboard	—	• Locked up	0.06
			• Mechanical failure	0.49
			• Contact failure	0.23
			• Connection/connector failure	0.22
13	Zener diode	—	• Short circuit	0.5
			• Open circuit	0.5
14	Small signal diode	—	• Parameter change	0.58
			• Short circuit	0.18
			• Open circuit	0.24
15	Triac	—	• Failed on	0.10
			• Failed off	0.90
16	Bipolar memory	—	• Data bit loss	0.21
			• Slow transfer of data	0.79

$$B_{si} = 12\, i^2 \left(\ln Z - \frac{D}{i} \right) \Big/ (6\, i + i + 1), \tag{12.1}$$

where

$$Z = \frac{1}{i} \sum_{K=1}^{i} T_K \tag{12.2}$$

$$D = \sum_{K=1}^{i} \ln T_K , \tag{12.3}$$

where

- i is the total number of failures contained in the sample.
- T_K is the kth time to failure.

Note that a sample of at least 20 failures is necessary for the test to discriminate effectively. If the times to failure are exponentially distributed, then B_{si} is distributed as chi-square with (i–1) degrees of freedom. Consequently, a two-tailed chi-square criterion is used.[39]

Example 12.1

A sample of 29 identical medical devices was tested, and the times to failure of these devices are presented in Table 12.10.

TABLE 12.10
Times to Failure of Medical Devices in Hours

4	36	84	143	269	353
5	46	88	182	279	363
24	47	109	183	280	365
23	62	110	264	351	435
35	63	142	265	352	

Determine if these times to failure can be represented by an exponential distribution with a 90% confidence level.

By inserting the given data into Equation (12.2), we get

$$Z = \frac{4962}{29} = 171.10 \text{ hours}.$$

Similarly, by using Table 12.10 data in Equation (12.3), we get

$$D = 134.5$$

Inserting the above calculated values and the given data into Equation (12.1) yields

$$B_{si} = 12(29)^2 \left(\ln 171.10 - \frac{134.5}{29} \right) \Big/ \{6(29) + 29 + 1\}$$

$$= 24.95 .$$

Using Table 9.1 for a two-tailed test with a 90% confidence level, the critical value of

$$\chi^2 \left[\frac{\theta}{2}, (i-1) \right] = \left[\frac{0.1}{2}, (29-1) \right] = 41.33 ,$$

where

$\theta = 1 -$ (confidence level) $= 1 - 0.90 = 0.1$

and

$$\chi^2 \left[\left(1 - \frac{\theta}{2}\right), (i-1) \right] = \chi^2 \left[\left(1 - \frac{0.1}{2}\right), (29-1) \right] = 16.92 .$$

The above three calculated values indicate that there is no contradiction to the assumption of exponential distribution.

12.6.2 Maximum Likelihood Estimation (MLE) Method

This is an excellent technique to estimate distribution parameters once the statistical distribution of the data under consideration is known either using test methods such as Bartlett or through other means.[42-44]

The method assumes that a random sample of say, times to failure $t_1, t_2, t_3, \ldots, t_n$ of size n is taken from a population with a probability density function, $f(t, \lambda)$, where t is time and λ the unknown parameter of the distribution. The joint density function of the n random variables of the random sample is called the likelihood function, and it is expressed by

$$L = f(t_1; \lambda)\, f(t_2; \lambda) \ldots f(t_n; \lambda) , \tag{12.4}$$

where

L is the maximum likelihood estimator (MLE) of λ.

Normally, λ is estimated by solving the equation

$$\frac{\partial \ln L}{\partial \lambda} = 0 . \tag{12.5}$$

Similarly, parameters more than one are estimated. More specifically, the values of parameters of distributions having more than one parameter.

The method is demonstrated through the following two examples:

Example 12.2

Assume that the times to failure of a pacemaker are described by the following probability density function:

$$f(t;\lambda) = \lambda e^{-\lambda t} \qquad t \geq 0, \tag{12.6}$$

where

$f(t;\lambda)$ is the failure density function.
t is time.
λ is the distribution parameter.

Develop an expression to estimate the value of λ by using the MLE method.

By inserting Equation (12.6) into Equation (12.4), we get

$$L = \lambda^n e^{-\lambda \sum_{i=1}^{n} t_i}. \tag{12.7}$$

Using Equation (12.7) in Equation (12.5) yields

$$\frac{\partial \ln L}{\partial \lambda} = \frac{n}{\lambda} - \sum_{i=1}^{n} t_i = 0. \tag{12.8}$$

By solving Equation (12.8), we get

$$\hat{\lambda} = \frac{n}{\sum_{i=1}^{n} t_i}, \tag{12.9}$$

where

$\hat{\lambda}$ is the estimate of λ.

Example 12.3

Assume that a sample of ten pacemakers was examined. The pacemaker times to failure are given in Table 12.11. Estimate pacemaker failure rate if the times to failure presented in Table 12.11 are exponentially distributed.

TABLE 12.11
Pacemaker Times to Failure in Hours

80,000	65,000
60,000	60,000
90,000	55,000
40,000	75,000
100,000	80,000

By substituting the specified data into Equation (12.9), we obtain

$$\hat{\lambda} = \frac{10}{705{,}000} = 1.4181 \times 10^{-5} \text{ failures/hour}.$$

It means the pacemaker failure rate is 1.4181×10^{-5} failures/hour.

12.7 PROBLEMS

1. What is a standard? Discuss its need.
2. Discuss the following organizations concerned with medical devices:
   - AAMI
   - FDA
   - ISO
3. Write an essay on MIL-HDBK-217.
4. Discuss briefly the responsibilities of the following U.S. bodies with respect to medical devices:
   - Consumer Product Safety Commission
   - Federal Communications Commission
   - Occupational Safety and Heath Administration
5. List at least five standards produced by the IEC with respect to medical devices/equipment.
6. Describe the following reliability documents with respect to their application to medical devices:
   - MIL-STD-338
   - MIL-STD-721C
   - MIL-STD-2155
7. Write an essay on failure data with respect to medical devices.
8. What are the typical sources for collecting failure data during a medical device/equipment life cycle?
9. Assume that the times to failure of a medical device are represented by the normal probability density function. Develop expressions to estimate the values of its parameters by using the MLE approach.

10. A sample of 20 identical medical devices was tested. The times to failure of these devices are given in Table 12.12. Determine if these times to failure can be represented by an exponential distribution with a 90% confidence level by using the Bartlett test.

TABLE 12.12
Times to Failure of Medical Devices in Hours

3	36	83	144
4	48	86	188
18	45	112	184
23	61	114	262
35	66	142	269

REFERENCES

1. Kanner, L., *History of Dentistry: Folklore of the Teeth*, Quintessence Publishing Company, Chicago, 1936, p. 205.
2. Hutt, P.B., A History of Government Regulation of Adulteration and Misbranding of Medical Devices, in *The Medical Device Industry*, N.F. Estrin, Ed., Marcel Dekker Inc., New York, 1990, pp. 17–33.
3. Darnton, R., *Mesmerism and the End of Enlightenment in France*, Harvard University Press, Cambridge, MA, 1968.
4. Fuller, R.C., *Mesmerism and the American Cure of Souls*, University of Pennsylvania Press, Philadelphia, 1982.
5. Janssen, W.F., The Gadgeteers, in *The Health Robbers*, S. Barrett and G. Knight, G.F., Eds., Stickley Co., Philadelphia, 1976, Chapter 16.
6. Bloch, G., *Mesmerism: A Translation of the Original Scientific and Medical Writings of F.A. Mesmer*, published by W. Kaufman, Los Altas, California, 1980.
7. Willingmyre, G.T., Industry's Role in Standards Development, in *The Medical Device Industry*, N.F. Estrin, Ed., Marcel Dekker Inc., New York, 1990, pp. 139–150.
8. Miller, M.J. and Bridgeman, E.A., Role of AAMI in the Voluntary Standards Process, in *The Medical Device Industry*, N.F. Estrin, Ed., Marcel Dekker Inc., New York, 1990.
9. Leitgeb, N., *Safety in Electromedical Technology*, Interpharm Press Inc., Buffalo Grove, IL, 1996.
10. Fries, R.C., *Reliability Assurance for Medical Devices, Equipment, and Software*, Interpharm Press Inc., Buffalo Grove, IL, 1991.
11. Appler, W.D. and McMann, G.L., Medical Device Regulation: The Big Picture, in *The Medical Device Industry*, N.F. Estrin, Ed., Marcel Dekker Inc., New York, 1990.
12. Fries, R.C., *Reliable Design of Medical Devices*, Marcel Dekker Inc., New York, 1997.
13. Hahn, R.F., Data Collection Techniques, *Proceedings of the Annual Reliability and Maintainability Symposium*, 1972, pp. 38–43.
14. Dhillon, B.S., *Design Reliability: Fundamentals and Applications*, CRC Press, Boca Raton, FL, 1999.
15. Government Industry Data Exchange Program (GIDEP), GIDEP Operations Center, Fleet Missile Systems, Analysis and Evaluation Group, Department of Defense, Corona, CA 91720, USA.

16. Parts Reliability Information Center (PRINCE), Reliability Office, George C. Marshall Space Flight Center, National Aeronautics and Space Administration (NASA), Huntsville, AL 35812, USA.
17. National Technical Information Service (NTIS), 5285 Port Royal Road, Springfield, VA 22161, USA.
18. Reliability Analysis Center (RAC), Rome Air Development Center (RADC), Griffiss Air Force Base, Department of Defense, Rome, NY 13441, USA.
19. Emergency Care Research Institute (ECRI), 5200 Butler Parkway, Plymouth Meeting, PA 19462, USA.
20. Center for Devices and Radiological Health, Food and Drug Administration (FDA), 1390 Piccard Drive, Rockville, MD 20850, USA.
21. Nobel, J.J., Role of ECRI, in the *Medical Device Industry*, N.F. Estrin, Ed., Marcel Dekker Inc., New York, 1990, pp. 177–198.
22. Arcarese, J.S., FDA's Role in Medical Device User Education, in the *Medical Device Industry*, N.F. Estrin, Ed., Marcel Dekker Inc., New York, 1990, pp. 129–138.
23. Grabarz, D.F. and Cole, M.F., Developing a Recall Program, in the *Medical Device Industry*, N.F. Estrin, Marcel Dekker Inc., New York, 1990, pp. 335–351.
24. Cooper, J.B., Newbower, R.S., and Kitz, R.J., An Analysis of Major Errors and Equipment Failures in Anaesthesia Management, *Anaesthesiology*, Vol. 60, 1984, pp. 34–42.
25. Cooper, J.B., Toward Prevention of Anaesthetic Mishaps, in *Analysis of Anaesthetic Mishaps*, E. Pierce and J. Cooper, Eds., International Anaesthesiology Clinics, Vol. 22, 1984, pp. 167–183.
26. Banta, H.D., The Regulation of Medical Devices, *Preventive Medicine*, Vol. 19, 1990, pp. 693–699.
27. U.S. Congress, House Committee on Interstate and Foreign Commerce: Medical Devices, Hearings before the Subcommittee on Public Health and the Environment, October 23–24, 1973, Serial No. 93-61, U.S. Government Printing Office, Washington, DC.
28. Micco, L.A., Motivation for the Biomedical Instrument Manufacturer, *Proceedings of the Annual Reliability and Maintainability Symposium*, 1972, pp. 242–244.
29. Dhillon, B.S., Reliability Technology in Health Care Systems, *Proceedings of the IASTED Int. Symp. Com. Adv. Technol. Med. Health Care Bioeng.*, 1990, pp. 84–87.
30. Schwartz, A.P., A Call for Real Added Value, *Medical Industry Executive*, February/March, 1994, pp. 5–9.
31. Clifton, B.S. and Hotten, W.I.T., Deaths Associated with Anaesthesia, *Br. J. Anaes.*, Vol. 35, 1963, pp. 250–259.
32. Dripps, R.D., Lamont, A., and Eckenhoff, J.F., The Role of Anaesthesia in Surgical Mortality, *JAMA*, Vol. 178, 1961, pp. 261–266.
33. Edwards, G., et al., Death Associated with Anaesthia: Report on 1000 Cases, *Anaesthesia*, Vol. 11, 1956, pp. 194–220.
34. Peterson, M.G.E., The Probability of Failure Depends on Who is Asking, *Proceedings of the 8th IEEE Symposium on Computer-Based Medical Systems*, 1995, pp. 51–56.
35. Barker, K.N. and McConnell, W.E., Detecting Errors in Hospitals, *Am. J. Hosp. Pharm.*, Vol. 19, August 1962, pp. 361–369.
36. MIL-HDBK-217F, Reliability Prediction of Electronic Equipment, Department of Defense, Washington, DC, USA.
37. Epstein, B., Tests for the Validity of the Assumption that the Underlying Distribution of Life is Exponential, *Technometrics*, Vol. 2, 1960, pp. 83–101.

38. Epstein, B., Tests for the Validity of the Assumption that the Underlying Distribution of Life is Exponential, *Technometrics*, Vol. 2, 1960, pp. 327–335.
39. Lamberson, L.R., An Evaluation and Comparison of Some Tests for the Validity of the Assumption that the Underlying Distribution of Life is Exponential, *AIIE Trans.*, Vol. 12, 1974, pp. 327–335.
40. Dhillon, B.S., *Quality Control, Reliability, and Engineering Design*, Marcel Dekker Inc., New York, 1985.
41. Dhillon, B.S., *Reliability Engineering in Systems Design and Operation*, Van Nostrand Reinhold Company, New York, 1983.
42. Shooman, M.L., *Probabilistic Reliability: An Engineering Approach*, McGraw-Hill Book Company, New York, 1968.
43. Lloyd, M. and Lipow, M., *Reliability: Management, Methods, and Mathematics*, Prentice-Hall Inc., Englewood Cliffs, NJ, 1961.
44. Mann, N., Shafer, R.E., and Singpurwalla, N.D., *Methods for Statistical Analysis of Reliability and Life Data*, John Wiley & Sons, Inc., New York, 1974.

Appendix

Bibliography: Literature on Medical Device Reliability and Associated Areas

A.1 INTRODUCTION

Over the years, many publications on medical device reliability and associated areas have appeared in the form of journal papers, conference proceeding papers, books, technical reports, and so on. This section presents an extensive list of these publications. The period covered by the listing is 1962 to 1999. The main objective of this listing is to provide readers with sources of additional information on medical device reliability and related areas.

A.2 PUBLICATIONS

1. A Computer Glitch Turns Miracle Machine into Monster for Three Cancer Patients, *People Weekly*, No. 26, November 24, 1986, pp. 48–50.
2. A Hospital Recall and Reporting System for Medical Devices, American Society for Hospital Engineering/American Hospital Association, Chicago, 1978.
3. AAMI MIM-3/84, Establishing and Administering Medical Instrumentation Maintenance Programs, Guideline for, Association for the Advancement of Medical Instrumentation (AAMI), Arlington, VA, 1984.
4. Al-Abdulla, H.M. and Lulu, D.J., Hypokalaemia and Pacemaker Failure, *American Surgery*, Vol. 40, 1974, pp. 234–236.
5. Allen, D., California Home to Almost One-Fifth of U.S. Device Industry, *Medical Device and Diagnostic Industry Magazine*,Vol. 19, No. 10, 1997, pp. 64–67.
6. Allen, R.C., FDA and the Cost of Health Care, *Medical Device and Diagnostic Industry Magazine*, Vol. 18, No. 7, 1996, pp. 28–35.
7. Anderson, F.A., Medical Device Risk Assessment, in *The Medical Device Industry*, edited by N.F. Estrin, Marcel Dekker Inc., New York, 1990, pp. 487–493.
8. Andrews, L.B., Stocking, C., Krizek, T., Gottlieb, L., Krizek, C., Vargish, T., and Siegler, M., *Lancet*, Vol. 349, 1997, pp. 309–313.

9. ANSI/AAMI HE-48, Human Factors Engineering Guidelines and Preferred Practices for the Design of Medical Devices, Association for the Advancement of Medical Instrumentation (AAMI), Arlington, VA, 1993.
10. Applegate, M.H., Diagnosis-Related Groups: Are Patients in Jeopardy?, in *Human Error in Medicine*, edited by M. Bogner, Lawrence Erlbaum Associates Publishers, Hillsdale, NJ, 1994, 349–371.
11. Appler, W.D. and McMann, G.L., Medical Device Regulation: The Big Picture, in *The Medical Device Industry*, edited by N.F. Estrin, Marcel Dekker Inc., New York, 1990, pp. 35–51.
12. Arcarese, J., An FDA Perspective, *Proceeedings of the First Symposium on Human Factors in Medical Devices*, 1989, pp. 23–24.
13. Arcarese, J.S., FDA's Role in Medical Device User Education, in *The Medical Device Industry*, edited by E.F. Estrin, Marcel Dekker Inc., 1990, pp. 129–138.
14. Baker, K.N and McConnell, W.E., Detecting Errors in Hospitals, *American Journal of Hospital Pharmacy*, Vol. 19, 1962, pp. 361–369.
15. Banta, H.D., The Regulation of Medical Devices, *Preventive Medicine*, Vol. 19, 1990, pp. 693–699.
16. Barker, K.N., Kimbrough, W.W., and Heller, W.M., *A Study of Medication Errors in a Hospital*, University of AR, Fayetteville, AR, 1966.
17. Barker, K.N. and McConnell, W.E., Detecting Errors in Hospitals, *American Journal of Hospitals*, Vol. 19, 1962, pp. 361–369.
18. Bassen, H., Silberberg, J., Houston, F., Knight, W., Christman, C., and Greberman, M., Computerized Medical Devices: Usage, Trends, Problems, and Safety Technology, *Proceedings of the 7th Annual Conference of the IEEE/Engineering in Medicine and Biology Society*, 1985, pp. 180–185.
19. Bassen, H., Silberberg, J., Houston, F., and Knight, W., Computerized Medical Devices, Trends, Problems, and Safety, *IEEE Aerospace and Electronic Systems (AES) Magazine*, September 1986, pp. 20–24.
20. Beach, J.E. and Bonewell, M.L., Setting–Up a Successful Software Vendor Evaluation/Qualification Process for "Off–the–Shelve" Commercial Software Used in Medical Devices, *Proceedings of the 6th Annual IEEE Symposium on Computer-Based Medical Systems*, 1993, pp. 284–288.
21. Beasley, L.J., Reliability and Medical Device Manufacturing, *Proceedings of the Annual Reliability and Maintainability Symposium*, 1995, pp. 128–131.
22. Beier, B., Liability and Responsibility for Clinical Medical Software in the Federal Republic of Germany, *Proceedings of the 10th Annual Symposium on Computer Applications Medical Care*, 1986, pp. 364–368.
23. Beith, B.H., Human Factors and the Future of Telemedicine, *Medical Device and Diagnostic Industry Magazine*, Vol. 21, No. 6, 1999, pp. 36–40.
24. Belkin, L., Human and Mechanical Failures Plague Medical Care, *The New York Times*, The Metro Section, March 21, 1992.
25. Bell, D.D., Contrasting the Medical-Device and Aerospace-Industries Approach to Reliability, *Proceedings of the Annual Reliability and Maintainability Symposium*, 1996, pp. 125–127.

26. Bethune, J., On Product Liability: Stupidity and Waste Abounding, *Medical Device and Diagnostic Industry Magazine*, Vol. 18, No. 8, 1996, pp. 8–11.
27. Bethune, J., The Cost–Effective Bugaboo, *Medical Device and Diagnostic Industry Magazine*, Vol. 19, No. 4, 1997, pp. 12–15.
28. Blum, L.L., Equipment Design and "Human" Limitations, *Anesthesiology*, Vol. 35, 1971, pp. 101–102.
29. Boardman, T.A., Dipasquale, T., Product Liability Implications of Regulatory Compliance or Noncompliance, in *The Medical Device Industry*, edited by N.F. Estrin, Marcel Dekker Inc., New York, 1990, pp. 387–397.
30. Bogner, M.S., Ed., *Human Error in Medicine*, Lawrence Erlbaum Associates Publishers Hillsdale, NJ, 1994.
31. Bogner, M.S., Human Error in Medicine: A Frontier for Change, in *Human Error in Medicine*, edited by M.S. Bogner, Lawrence Erlbaum Associates Publishers, Hillsdale, NJ, 1994, pp. 373–383.
32. Bogner, S., Medical Devices: A New Frontier for Human Factors, *CSERIAC Gateway*, Vol. IV, No. 1, 1993, pp. 12–14.
33. Boumil, M.M. and Elias, C.E., *The Law of Medical Liability in a Nutshell*, West Publishing Co., St. Paul, MN, 1995.
34. Bousvaros, G.A., Don, C., and Hopps, J.A., An Electrical Hazard of Selective Angiocardiography, *Canadian Medical Association Journal*, Vol. 87, 1962, pp. 286–288.
35. Bracco, D., How to Implement a Statistical Process Control Program, *Medical Device and Diagnostic Industry Magazine*, Vol. 3, No. 3, 1998, pp. 129–134.
36. Brennan, T.A., Leape, L.L., Laird, N.M., Hebert, L., Localio, A.R., Lawthers, A.G., Newhouse, J.P., Weiler, P.C., and Hiatt, H.H., Incidence of Adverse Events and Negligence in Hospitalized Patients, *The New England Journal of Medicine*, Vol. 324, 1991, pp. 370–376.
37. Brennan, T.A., Localio, A.R., and Laird, N.M., Reliability and Validity of Judgements Concerning Adverse Events and Negligence, *Med. Care*, Vol. 27, 1989, pp. 1148–1158.
38. Brown, S.L., Use of Risk Assessment Procedures for Evaluating Risks of Ethylene Oxide Residues in Medical Devices, in *The Medical Device Industry*, edited by N.F. Estrin, Marcel Dekker Inc., New York, 1990, pp. 469–485.
39. Bruley, M.E., Ergonomics and Error–Who is Responsible?, *Proceedings of the First Symposium on Human Factors in Medical Devices*, 1989, pp. 6–10.
40. Bruner, J.M.R., Hazards of Electrical Apparatus, *Anesthesiology*, Vol. 28, No.2, 1967, pp. 396–425.
41. Burchell, H.B., Electrocution Hazards in the Hospital or Laboratory, *Circulation*, 1963, pp. 1015–1017.
42. Cady, W.W. and Iampietro, D., Medical Device Reporting, *Medical Device and Diagnostic Industry Magazine*, Vol. 18, No. 5, 1996, pp. 58–67.
43. Camishion, R.C., Electrical Hazards in the Research Laboratory, *Journal of Surgical Research*, Vol. 6, 1966, pp. 221–227.
44. Casey, S., *Set Phasers on Stun*, Aegean Publishing Company, Santa Barbara, CA, 1998.

45. Chevlin, D.H. and Jorgens, J., Medical Device Software Requirements: Definition and Specification, *Medical Instrumentation*, Vol. 30, No. 2, March/April 1996.
46. Clayton, M., The Right Way to Prevent Medication Errors, *Registered Nurse (RN)*, June 1987, pp. 30–31.
47. Cohen, T., Computerized Maintenance Management Systems: How to March Your Department's Needs with Commercially Available Products, *Journal of Clinical Engineering*, November/December, 1995, pp. 457–461.
48. Cohen, T., Validating Medical Equipment Repair and Maintenance Metrics: A Progress Report, *Biomedical Instrumentation and Technology*, January/February 1977, pp. 23–32.
49. Cook, R.I., Woods, D.D., Operating at the Sharp End: The Complexity of Human Error, in *Human Error in Medicine*, edited by M.S. Bogner, Lawrence Erlbaum Associates, Publishers, Hillsdale, NJ, 1994, pp. 255–310.
50. Cooper, J.B., Newbower, R.S., Long, C.D., McPeek, B., Preventable Anesthetic Mishaps: A Study of Human Factors, *Anesthesiology*, Vol. 49, 1978, pp. 399–406.
51. Cooper, J.B., Newbower, R.S., Ritz, R.J., An Analysis of Major Errors and Equipment Failures in Anesthesia Management: Considerations for Prevention and Detection, *Anesthesiology*, Vol. 60, 1984, pp. 34–42.
52. Couch, N.P., Tilney, N.L., Rayner, A.A., and Moore, F.D., The High Cost of Low Frequency Events, *The New England Journal of Medicine*, Vol. 304, No. 11, 1981, pp. 634–637.
53. Dawson, N.V., Systematic Errors in Medical Decision Making: Judgement Limitations, *Journal of General Internal Medicine*, Vol. 2, 1987, pp. 183–187.
54. *Device Good Manufacturing Practices Manual*, Center for Devices and Radiological Health, Department of Health and Human Services, Washington, DC, 1987.
55. *Device Recalls: A Study of Quality Problems*, Food and Drug Administration, Rockville, MD, 1990.
56. Dhillon, B.S., Bibliography of Literature on Medical Equipment Reliability, *Microelectronics and Reliability*, Vol. 20, 1980, pp. 737–742.
57. Dhillon, B.S., *Design Reliability: Fundamentals and Applications*, CRC Press, Boca Raton, FL, 1999.
58. Dhillon, B.S., *Reliability Engineering Applications: Bibliography on Important Application Areas*, Beta Publishers Inc., Gloucester, Ontario, 1992, Chapter 13.
59. Dhillon, B.S., *Reliability Engineering in Systems Design and Operation*, Van Nostrand Reinhold Company, New York, 1983, Chapter 11.
60. Dhillon, B.S., Reliability Technology in Health Care Systems, *Proceedings of the IASTED International Symposium on Comp. Adv. Technol. Med. Health Care Bio-eng.*, 1990, pp. 84–87.
61. Dhillon, B.S., Tools for Improving Medical Equipment Reliability and Safety, *Physics in Medicine and Biology*, Vol. 39a, 1994, pp. 941.

62. Dickson, C., World Medical Electronics Market: An Overview, *Medical Devices and Diagnostic Industry Magazine*, Vol. 6, No. 5, 1984, pp. 53–58.
63. Donchin, Y., Biesky, M., Cotev, S., Gopher, D., Olin, M., Badihi, Y., and Cohen, G., The Nature and Causes of Human Errors in a Medical Intensive Care Unit, *Proceedings of the 33rd Annual Meeting of the Human Factors Society*, 1989, pp. 111–118.
64. Donohue, J. and Apostolou, S.F., Predicting Shelf Life from Accelerated Aging Data: The D & A and Variable x10 Techniques, *Medical Device and Diagnostic Industry Magazine*, Vol. 20, No. 6, 1998, pp. 68–72.
65. Donohue, J. and Apostolou, S.F., Shelf-Life Prediction for Radiation-Sterilized Plastic Devices, *Medical Device and Diagnostic Industry Magazine*, Vol. 12, No. 1, 1990, pp. 124–129.
66. Donohue, J., Predicting Post-Rad Shelf Life From Accelerated Aging Data: The D & A Process, *Technical Papers*, Society of Plastics Engineers, Brookfield, CT, Vol. 42, 1996, pp. 2819–2822.
67. Drefs, M.J., IEC 601-1 Electrical Safety Testing, *Compliance Engineering*, Vol. XIV, No.2, March/April 1997.
68. Drefs, M.J., IEC 601-1 Electrical Testing, Part 2: Performing the Tests, *Compliance Engineering*, Vol. XIV, No. 3, May/June, 1997.
69. Drury, C.G., Schiro, S.C., Czaja, S.J., and Barnes, R.E., Human Reliability in Emergency Medical Response, *Proceedings of the Annual Reliability and Maintainability Symposium*,1977, pp. 38–42.
70. Dubois, R.W. and Brook, R.H., Preventable Deaths: Who, How, and Why?, *Annals of Internal Medicine*, Vol. 109, 1988, pp. 582–589.
71. Eberhard, D.P., Qualification of High Reliability Medical Grade Batteries, *Proceedings of the Annual Reliability and Maintainability Symposium*, 1989, pp. 356–362.
72. Edwards, F.V., Before Design: Thoroughly Evaluate Your Concept, *Medical Device and Diagnostic Industry Magazine*, Vol. 19, No.3, 1997, pp. 46–50.
73. Egeberg, R.O., Engineers and the Medical Crisis, *Proceedings of the IEEE*, Vol. 57, No. 11, 1969, pp. 1807–1808.
74. Eichhorn, J.H., Prevention of Intra operative Anesthesia Accidents and Related Severe Injury Through Safety Monitoring, *Anesthesiology*, Vol. 70, 1989, pp. 572–577.
75. Eisenberg, P., Computer/User Interface Design Specification for Medical Devices, *Proceedings of the 6th Annual IEEE Symposium on Computer-Based Medical Systems*, 1993, pp. 177–182.
76. Elahi, B.J., Safety and Hazard Analysis for Software-Controlled Medical Devices, *Proceedings of the 6th Annual IEEE Symposium on Computer-Based Medical Systems*, 1993, pp. 10–15.
77. Elliott, L. and Mojdehbakhsh, R., A Process for Developing Safe Software, *Proceedings of the 7th Symposium on Computer-Based Medical Systems*, 1994, pp. 241–246.
78. *Essential Standards for Biomedical Equipment Safety*, Association for the Advancement of Medical Instrumentation, Arlington, VA, 1985.

79. Estrin, N.F., Ed., *The Medical Device Industry*, Marcel Dekker Inc., New York, 1990.
80. Fairhurst, G.A. and Murphy, K.L., Help Wanted, *Proceedings of the Annual Reliability and Maintainability Symposium*, 1976, pp. 103–106.
81. Fatal Radiation Dose in Therapy Attributed to Computer Mistake, *The New York Times*, June 21, 1986, pp. 50.
82. *FDA Device Inspections Manual*, Food and Drug Administration, Arlington, VA, 1994.
83. Finfer, S.R., Pacemaker Failure on Induction of Anaethesia, *British Journal of Anaesthesia*, Vol. 66, 1991, pp. 509–512.
84. Finger, T.A., The Alpha System, *Proceedings of the Annual Reliability and Maintainability Symposium*, 1976, pp. 92–96.
85. Freihrr, G., Safety is Key to Product Quality, Productivity, *Medical Device and Diagnostic Industry Magazine*, Vol. 19, No. 4, 1997, pp. 18–19.
86. Fries, R., Human Factors and System Reliability, *Medical Device Technology*, March 1992, pp. 42–46.
87. Fries, R.C., *Medical Device Quality Assurance and Regulatory Compliance*, Marcel Dekker Inc., New York, 1998.
88. Fries, R.C., Pienkowski, P., and Jorgens, J., Safe, Effective and Reliable Software Design and Development for Medical Devices, *Medical Instrumentation*, Vol. 30, No. 2 1996, pp. 75– 80.
89. Fries, R.C., *Reliability Assurance for Medical Devices, Equipment, and Software*, Interpharm Press, Buffalo Grove, IL, 1991.
90. Fries, R.C., *Reliable Design of Medical Devices*, Marcel Dekker Inc., New York, 1997.
91. Fries, R.C., Willingmyre, G.T., Simons, D., and Schwartz, R.T., Software Regulation, in *The Medical Device Industry*, edited by N.F. Estrin, Marcel Dekker Inc., New York, 1990, pp. 557–569.
92. Gaba, D.M., Human Error in Anesthetic Mishaps, *International Anesthesiology Clinics*, Vol. 27, No. 3, 1989, pp. 137–147.
93. Gaba, D.M., Human Error in Dynamic Medical Domains, in *Human Error in Medicine*, edited by M.S. Bogner, Lawrence Erlbaum Associates Publishers, Hillsdale, NJ, 1994, pp. 197–224.
94. Gaba, D.M., Maxwell, M., and DeAnda, A., Anesthetic Mishaps: Breaking the Chain of Accident Evolution, *Anesthesiology*, Vol. 66, 1987, pp. 670–676.
95. Galer, I.A. and Yap, B.L., Ergonomics in Intensive Care, Applying Human Factors Data to the Design and Evaluation of Patient Monitoring Systems, *Ergonomics*, Vol. 23, 1980, pp. 763–779.
96. Gay, R.J. and Brown, D.F., Pacemaker Failure Due to Procainamide Toxicity, *American Journal of Cardiology*, Vol. 34, 1974, pp. 728–732.
97. Gechman, R., Tiny Flaws in Medical Design Can Kill, *Hosp. Top.*, Vol. 46, 1968, pp. 23– 24.
98. Gingerich, D., Ed., *Medical Product Liability*, F&S Press, New York, 1981.
99. Gleason, K.L., Precautionary Checklist for Addressing the Issue of Medical Device User Error and Product Misuse, in *The Medical Device Industry*, edited E.F. Estrin, by Marcel Dekker Inc., New York, 1990, pp. 113–128.

100. Gopher, D., Olin, M., Donchin, Y., Bieski, M., Badihi, Y, Cohen, G., and Cotev, S., The Nature and Causes of Human Errors in a Medical Intensive Care Unit, *Proceedings of the Human Factors Society 33rd Annual Meeting*, 1989, pp. 956–960.
101. Gosbee, J. and Ritchie, E.M., Human Computer Interaction and Medical Software Development, *Interactions*, Vol. 4, No. 4, 1997, pp. 13–18.
102. Gosbee, J., The Discovery Phase of Medical Device Design: A Blend of Intuition, Creativity and Science, *Medical Device and Diagnostic Industry Magazine*, Vol. 19, No. 11, 1997, pp. 79–85.
103. Gowen, L.D., Specifying and Verifying Safety-Critical Software Systems, *Proceedings of the IEEE 7th Symposium on Computer-Based Medical Systems*, 1994, pp. 235–240.
104. Gowen, L.D. and Yap, M.Y., Traditional Software Development's Effects on Safety, *Proceedings of the 6th Annual IEEE Symposium on Computer-Based Medical Systems*, 1993, pp. 58–63.
105. Grabarz, D.F. and Cole, M..F., Developing a Recall Program, in *The Medical Device Industry*, edited by N.F. Estrin, Marcel Dekker Inc., New York, 1990 pp. 335–351.
106. Grant, L.J., Product Liability Aspects of Bioengineering, *Journal of Biomedical Engineering*, Vol. 12, 1990, pp. 262–266.
107. Grant, L.J., Regulations and Safety in Medical Equipment Design, *Anaesthesia*, Vol. 53, 1998, pp. 1–3.
108. Greatbatch, W., Designing for High Reliability in Medical Electronic Equipment, in *Medical Engineering*, edited by C. Ray, Yearbook Medical Publishers, 1972, p. 1003.
109. Gruppen, L.D., Wolf, F.M., and Billi, J.E., Information Gathering and Integration as Sources of Error in Diagnostic Decision Making, *Medical Decision Making*, Vol. 11, No. 4, 1991, pp. 233–239.
110. Gruppen, L.D., Wolf, F.M., and Billi, J.E., Information Gathering and Integration as Sources of Error in Diagnostic Decision Making, *Medical Decision Making*, Vol. 11, No. 4, 1991, pp. 233–239.
111. Hales, R.F., Quality Function Deployment in Concurrent Product/Process Development, *Proceedings of the 6th Annual IEEE Symposium on Computer-Based Medical Systems*, 1993, pp. 28–33
112. Harrison, D.W. and Kelly, P.L., Home Health Care: Accuracy, Calibration, Exhaust, and Failure Rate Comparisons of Digital Blood Pressure Monitors, *Med. Instrum.*, Vol. 21, No. 6, 1987, pp. 323–328.
113. Haslam, K.R. and Bruner, J.M.R., The Epidemiology of Failure in Cardiac Monitoring Systems, *Med. Instrum.*, Vol. 7, 1973, pp. 293–296.
114. Helmreich, R.L. and Schaefer, H.G., Team Performance in the Operating Room, in *Human Error in Medicine*, edited by M.S. Bogner, Lawrence Erlbaum Associates, Publishers, Hillsdale, NJ, 1994, pp. 225–253.
115. Heydrick, L., Jones, K.A., Applying Reliability Engineering During Product Development, *Medical Device and Diagnostic Industry Magazine*, Vol. 18, No. 4, 1996, pp. 80–84.

116. Holstein, H.M., *Inspection of Medical Device Manufacturers*, Health Industry Manufacturers Association, Washington, DC, 1983.
117. Hooten, W.F., A Brief History of FDA Good Manufacturing Practices, *Medical Device and Diagnostic Industry Magazine*, Vol. 18, No.5, 1996, p. 96.
118. Hopps, J.A., Electrical Hazards in Hospital Instrumentation, *Proceedings of the Annual Symposium on Reliability*, 1969, pp. 303–307.
119. Hough, G.W., *Preproduction Quality Assurance for Healthcare Manufacturers*, Interpharm Press, Buffalo Grove, IL, 1997.
120. Hutt, P.B., A History of Government Regulation of Adulteration and Misbranding of Medical Devices, in *The Medical Device Industry*, edited by N.F. Estrin, Marcel Dekker Inc., New York, 1990, pp. 17–33.
121. Hyman, W.A., An Evaluation of Recent Medical Device Recalls, *Medical Device and Diagnostic Industry Magazine*, Vol. 4, No. 11, 1982, pp. 53–55.
122. Hyman, W.A., Errors in the Use of Medical Equipment, in *Human Error in Medicine*, edited by M.S. Bogner, Lawrence Erlbaum Associates Publishers, Hillsdale, NJ, 1994, pp. 327–347.
123. Hyman, W.A., Human Factors in Medical Devices, in *Encyclopedia of Medical Devices*, edited by J.G. Webster, John Wiley & Sons, Inc., New York, 1988, pp. 1542–1553.
124. IEC 601-1: Safety of Medical Electrical Equipment, Part l: General Requirements, International Electrotechnical Commission (IEC), Geneva, 1977.
125. IEEE Case Studies in Medical Instrument Design, Institute of Electrical and Electronics Engineers, Piscataway, NJ, 1992.
126. Isaacson, L.H. and Taylor, E.F., SpecificationX-1414: A Reliability Milestone, *Proceedings of the Annual Symposium on Reliability*, 1970, pp. 86–91
127. ISO/DIS 14971, Medical Devices-Risk Management-Part I: Application of Risk Analysis to Medical Devices, International Organization for Standardization (ISO), Geneva, Switzerland, 1996.
128. Jacky, J., Risks in Medical Electronics, *Communications of the ACM*, Vol. 33, No. 12, 1990, p. 138.
129. Johns, R.J., What is Blunting the Impact of Engineering on Hospitals?, *Proceedings of the IEEE*, Vol. 57, No. 11, 1969, pp. 1823–1827.
130. Johnson, J.P., Reliability of ECG Instrumentation in a Hospital, *Proceedings of the Annual Symposium on Reliability*, 1969, pp. 314–318.
131. Johnson, W.G., Brennan, T.A., Newhouse, J.P., Leape, L.L., Lawthers, A.G., Hiath, H.H., and Weiler, P.C., The Economic Consequences of Medical Injuries, *JAMA*, Vol. 267, 1992, pp. 2487–2492.
132. Jorgens, J., Computer Hardware and Software as Medical Devices, *Medical Device and Diagnostic Industry Magazine*, Vol. 5, No. 5, 1983, pp. 62–67.
133. Jorgens, J., The Purpose of Software Quality Assurance: A Means to an End, in *Developing Safe, Effective, Reliable Medical Software*, Association for the Advancement of Medical Instrumentation, Arlington, VA, 1991, pp. 1–6.
134. Joyce, E., Firms Warns of Another Therc-20 Problem, *American Medical News*, November 7, 1986, pp. 20–21.

135. Joyce, E., Malfunction 54: Unraveling Deadly Medical Mystery of Computerized Accelerator Gone Awry, *American Medical News*, October 3, 1986, pp. 1–8.
136. Joyce, E., Software Bug Discovered in Second Linear Accelerator, *American Medical News*, November 7, 1986, pp. 20–21.
137. Joyce, E., Software Bugs: a Matter of Life and Liability, *Datamation*, 1987, Vol. 33, No. 10, 1987, pp. 88–92.
138. Kagey, K.S., Reliability in Hospital Instrumentation, *Proceedings of the Annual Reliability and Maintainability Symposium*, 1973, pp. 85–88.
139. Kandel, G.L. and Ostrander, L.E., Preparation of the Medical Equipment Designer: Academic Opportunities and Constraints, *Proceedings of the First Symposium on Human Factors in Medical Devices*, 1989, pp. 11–13.
140. Karp, D., Your Medication Orders Can Become Malpractice Traps, *Medical Economics*, February 1, 1988, pp. 79–91.
141. Keller, A.Z., Kamath, A.R.R., and Peacock, S.T., A Proposed Methodology for Assessment of Reliability, Maintainability and Availability of Medical Equipment, *Reliability Engineering*, Vol. 9, 1984, pp. 153–174.
142. Kim, J.S., Determining Sample Size for Testing Equivalence, *Medical Device and Diagnostic Industry Magazine*, Vol. 19, No. 5, 1997, pp. 114–117.
143. Kim, J.S. and Larsen, M., Improving Quality with Integrated Statistical Tools, *Medical Device and Diagnostic Industry Magazine*, Vol. 18, No. 10, 1996, pp.78–82.
144. Klatzky, R.L., Geiwitz, J., and Fischer, S., Using Statistics in Clinical Practice: A Gap Between Training and Application, in *Human Error in Medicine*, edited by M.S. Bogner, Lawrence Erlbaum Associates, Publishers, Hillsdale, NJ, 1994, pp. 123–140.
145. Knepell, P.L., Integrating Risk Management with Design Control, *Medical Device and Diagnostic Industry Magazine*, Vol. 20, No. 10, 1998, pp. 83–89.
146. Knight, J.C., Issues of Software Reliability in Medical Systems, *Proceedings of the 3rd Annual IEEE Symposium on Computer-Based Medical Systems*, 1990, pp. 124–129.
147. Kortstra, J.R.A., Designing for the User, *Medical Device Technology*, January/February 1995, pp. 22–28.
148. Kriewall, T.J. and Widin, G.P., An Application of Quality Function Deployment to Medical Device Development, in *Case Studies in Medical Instrument Design*, The Institute of Electrical and Electronics Engineers Inc., New York, 1991.
149. Krishnamurty, G.B., Community Health Education Programs, *Proceedings of the Annual Reliability and Maintainability Symposium*, 1976, pp. 97–102.
150. Krueger, G.P., Fatigue, Performance, and Medical Error, in *Human Error in Medicine*, edited by M.S. Bogner, Lawrence Erlbaum Associates, Publishers, Hillsdale, NJ, 1994, pp. 311–326.
151. Kuwahara, S.S., *Quality Systems and GMP Regulations for Device Manufacturers*, Interpharm Press, Buffalo Grove, IL, 1998.

152. LaBudde, E.V., Design Controls, *Medical Device and Diagnostic Industry Magazine*, Vol. 19, No. 6, 1997, pp. 40–45.
153. Landoll, J.R. and Caceres, C.A., Automation of Data Acquisition in Patient Testing, *Proceedings of the IEEE*, Vol. 57, No. 11, 1969, p. 100.
154. Leape, L.L., Error in Medicine, *JAMA*, Vol. 272, No. 23, 1994, pp. 1851–1857.
155. Leape, L.L., The Preventability of Medical Injury, in *Human Error in Medicine*, edited by M.S. Bogner, Lawrence Erlbaum Associates, Publishers, Hillsdale, NJ, 1994, pp. 13–25.
156. LeCocq, A.D., Application of Human Factors Engineering in Medical Product Design, *J. Clin. Eng.*, Vol. 12, No. 4, 1987, pp. 271–277.
157. Ledley, R.S., Practical Problems in the Use of Computers in Medical Diagnosis, *Proceedings of the IEEE*, Vol. 57, No. 11, 1969, p. 1941.
158. Leffingwell, D.A. and Norman, B., Software Quality in Medical Devices: A Top–Down Approach, *Proceedings of the 6th Annual IEEE Symposium on Computer-Based Medical Systems*, 1993, pp. 307–311.
159. Leitgeb, N., *Safety in Electromedical Technology*, Interpharm Press, Buffalo Grove, IL, 1996.
160. Levin, M., Human Factors in Medical Devices: A Clear and Present Danger, *Proceedings of the First Symposium on Human Factors in Medical Devices*, 1989, pp. 28–29.
161. Levkoff, B., Increasing Safety in Medical Device Software, *Medical Device and Diagnostic Industry Magazine*, Vol. 18, No. 9, 1996, pp. 92–97.
162. Linberg, K.R., Defining the Role of Software Quality Assurance in a Medical Device Company, *Proceedings of the 6th Annual IEEE Symposium on Computer-Based Medical Systems*, 1993, pp. 278–283.
163. Link, D.M., Current Regulatory Aspects of Medical Devices, *Proceedings of the Annual Reliability and Maintainability Syposium*, 1972, pp. 249–250.
164. Maddox, M.E., Designing Medical Devices to Minimize Human Error, *Medical Device and Diagnostic Magazine*, Vol. 19, No. 5, 1997, pp. 166–180.
165. Manthei, R.D., Organizing for Compliance: Reflections Twenty Years Later, *Medical Device and Diagnostic Industry Magazine*, Vol. 21, No. 6, 1999, pp. 103–105.
166. Maschino, S. and Duffell, W., Can Inspection Time be Reduced? Developing a HACCP Plan, *Medical Device and Diagnostic Industry Magazine*, Vol. 20, No. 10, 1998, pp. 64–68.
167. Mazur, G., Quality Function Development for a Medical Device, *Proceedings of the 6th Annual IEEE Symposium on Computer-Based Medical Systems*, 1993, pp. 22–27.
168. Mccune, C.G., Development of an Automated Pacemaker Testing System, *Proceedings of the 27th Midwest Symposium on Circuits and Systems*, 1984, pp. 9–12.
169. McDonald, J.S., Peterson, S., Lethal Errors in Anesthesiology, *Anethesiology*, Vol. 63, 1985, pp. A 497.
170. McLean, H., Exceeding the limits of Traditional Reliability Tests, *Medical Device and Diagnostic Industry Magazine*, Vol. 16, No. 4, 1994, pp. 96–100.

171. McLinn, J.A., Reliability Development and Improvement of a Medical Instrument, *Proceedings of the Annual Reliability and Maintainability Symposium*, 1996, pp. 236–242.
172. Mead, D., Human Factors Evaluation of Medical Devices, *Proceedings of the First Symposium on Human Factors in Medical Devices*, 1989, pp. 17–18.
173. Meadows, S., Human Factors Issues with Home Care Devices, *Proceedings of the First Symposium on Human Factors in Medical Devices*, 1989, pp. 37–38.
174. Medical Device Reporting for Manufacturers, Food and Drug Administration, Department of Health and Human Services, Washington, DC, 1996.
175. Medical Devices; Current Good Manufacturing Practice (GMP) Final Rule; Quality System Regulation, Food and Drug Administration, Department of Health and Human Services,Washington, DC, 1996.
176. Medication Errors Linked to Heavy Pharmacy Workloads, *Drug Topics*, December 10, 1990, pp. 12.
177. Meyer, J.L., Some Instrument Induced Errors in the Electrocardiogram, *JAMA*, Vol. 201, 1967, pp. 351–358.
178. Micco, L.A., Motivation for the Biomedical Instrument Manufacturer, *Proceedings of the Annual Reliability and Maintainability Symposium*, 1972, pp. 242–244.
179. Miller, M.J. and Bridgeman, E.A., Role of AAMI in the Voluntary Standards Process, in *The Medical Device Industry*, edited by N.F. Estrin, Marcel Dekker Inc., New York, 1990, pp. 163–175.
180. Mojdehbakhsh, R., Tsai, W.T., Kirani, S., and Elliott, L., Retrofitting Software Safety in an Implantable Medical Device, *IEEE Software*, No. 1, January 1994, pp. 41–50.
181. Montanez, J., *Medical Device Quality Assurance Manual*, Interpharm Press, Buffalo Grove, IL, 1996.
182. Mosenkis, R., Critical Alarms, *Proceedings of the First Symposium on Human Factors in Medical Devices*, 1989, pp. 25–27.
183. Mosenkis, R., Human Factors Issues with Critical Care, Respiratory Care, and Emergency Care Devices, *Proceedings of the First Symposium on Human Factors in Medical Devices*, 1989, p. 36.
184. Murray, K., Canada's Medical Device Industry Faces Cost Pressures, Regulatory Reform, *Medical Device and Diagnostic Industry Magazine*, Vol. 19, No. 8, 1997, pp. 30–39.
185. Neumann, P.G., Illustrative Risks to the Public in the Use of Computer Systems and Related Technology: Update on Therac-25, *ACM SIGSOFT Software Engineering Notes*, Vol. 12, No. 3, 1987, p. 3.
186. Neumann, P.G., Some Computer-Related Disasters and Other Egregious Horrors, *Proceedings of the 7th Annual Conference of the IEEE/Engineering in Medicine and Biology Society*, 1985, pp. 1238–1239.
187. Nevland, J.G., Electrical Shock and Reliability Considerations in Clinical Instruments, *Proceedings of the Annual Symposium on Reliability*, 1969, pp. 308–313.

188. Nichols, T.R., Dummer, S., Assessing Pass/Fail Testing When There Are No Failures to Assess, *Medical Device and Diagnostic Industry Magazine*, Vol. 19, No. 6, 1997, pp. 97–100.
189. Nobel, J.J., Human Factors Design of Medical Devices: The Current Challenge, *Proceedings of the First Symposium on Human Factors in Medical Devices*, 1989, pp. 1–5.
190. Nobel, J.J., Medical Device Failures and Adverse Effects, *Pediatric Emergency Care*, Vol. 7, No. 2, 1991, pp. 120–123.
191. Nobel, J.J., Role of ECRI, in *The Medical Device Industry*, edited by N.F. Estrin, Marcel Dekker Inc., New York, 1990, pp. 117–198.
192. Norman, J.C. and Goodman, L., Acquaintance with and Maintenance of Biomedical Instrumentation, *J. Assoc. Advan. Med. Inst.*, September 1966, pp. 8–9.
193. Northrup, S.J., Reprocessing Single-Use Devices: An Undue Risk, *Medical Device and Diagnostic Industry Magazine*, Vol. 21, No. 3, 1999, pp. 38–41.
194. O'Leary, D.J., International Standards: Their New role in a Global Economy, *Proceedings of the Annual Reliability and Maintainability Symposium*, 1996, pp. 17–23.
195. O'Reilly, M.V., Murnaghan, D.P., Williams, M.B., Transvenous Pacemaker Failure Induced by Hyperkalaemia, *JAMA*, Vol. 228, 1974, pp. 336–337.
196. Olivier, D.P., Engineering Process Improvement Through Error Analysis, *Medical Device and Diagnostic Industry Magazine*, Vol. 21, No. 3, 1999, pp. 130–136.
197. Olivier, D.P., Software Safety: Historical Problems and Proposed Solutions, *Medical Device and Diagnostic Industry Magazine*, Vol. 17, No. 7, 1995, pp. 116–124.
198. Onel, S., Draft Revision of FDA's Medical Device Software Policy Raises Warning Flags, *Medical Device and Diagnostic Industry Magazine*, Vol. 19, No. 10, 1997, pp. 82–91.
199. Ozog, H., Risk Management in Medical Device Design, *Medical Device and Diagnostic Industry Magazine*, Vol. 19, No. 10, 1997, pp. 112–115.
200. Pelnik, T.M and Suddarth, G.J., Implementing Training Programs for Software Quality Assurance Engineers, *Medical Device and Diagnostic Industry Magazine*, Vol. 20, No. 10, 1998, pp. 75–85.
201. Perper, J.A., Life-Threatening and Fatal Therapeutic Misadvantages, in *Human Error in Medicine*, edited by M.S. Bogner, Lawrence Erlbaum Associates Publishers, Hillsdale, NJ, 1994, pp. 27–52.
202. Peterson, M.G.E., The Probability of Failure Depends on Who is Asking, *Proceedings of the 8th IEEE Symposium on Computer Based-Medical Systems*, 1995, pp. 51–56.
203. Philip, J.H., Human Factors in the Anesthesia Workplace: A User's View, *Proceedings of the First Symposium on Human Factors in Medical Devices*, 1989, pp. 19–22.
204. Pickett, R.M., Triggs, T.J., *Human Factors in Health Care*, Lexington Books, DC Heath and Company, Lexington, MA, 1975.

205. Plamer, B., Managing Software Risk, *Medical Device and Diagnostic Industry Magazine*, Vol. 19, No. 9, 1997, pp. 36–40.
206. Preproduction Quality Assurance Planning Recommendations for Medical Device Manufacturers, Compliance Guidance Series, HHS Publication FDA 90–4236, Food and Drug Administration (FDA), Rockville, MD, September 1989.
207. *Proceedings of the First Symposium on Human Factors in Medical Devices,* Plymoth Meeting, Pennsylvania, December 1989.
208. Purday, J.P. and Towey, R.M., Apparent Pacemaker Failure Caused by Activation of Ventricular Threshold Test by a Magnetic Instrument Mat During General Anaesthesia, *British Journal of Anaesthesia*, Vol. 69, 1992, pp. 645–646.
209. *Quality System Regulation*, Food and Drug Administration (FDA), Department of Health and Human Services, Washington, DC, 1996.
210. Raddalgoda, M., Zero Downtime: Setting the Standard for Reliability with Microkernel RTOS Technology, *Medical Device and Diagnostic Industry Magagine*, Vol. 20, No. 10, 1998, pp. 100–105.
211. Rappaport, M., Human Factors Applications in Medicine, *Human Factors*, Vol. 12, No. 1, 1970, pp. 25–35.
212. Rastogi, A.K., High-Quality Care at Low Cost, *Medical Device and Diagnostic Industry Magazine*, Vol. 19, No. 7, 1997, pp. 28–32.
213. Rau, G. and Tripsel, S., Ergonomic Design Aspects in Interaction Between Man and Technical Systems in Medicine, *Med. Prog. Technol.*, Vol. 9, 1982, pp. 153–159.
214. Redig, G. and Swanson, M., Total Quality Management for Software Development, *Proceedings of the 6th Annual IEEE Symposium on Computer-Based Medical Systems*, 1993, pp. 301–306.
215. Reeter, A.K., The Role of Training in Human Factors, *Proceedings of the First Symposium on Human Factors in Medical Devices*, 1989, pp. 30–31.
216. Reliability Technology for Cardiac Pacemakers, NBS/FDA Workshop, June 1974, NBS Publication No. 400-28, National Bureau of Standards, Department of Commerce, Washington, DC.
217. Reupke, W.A., Srinivasan, R., Rigterink, P.V., and Card, D.N., The Need for a Rigorous Development andTesting Methodology for Medical Software, *Proceedings of the Symposium on the Engineering of Computer-Based Medical Systems*, 1988, pp. 15–20.
218. Rice, J.A., The Impact of Cost Control in Restructuring of the Health Care Industry, in *The Medical Device Industry*, edited by N.F. Estrin, Marcel Dekker Inc., New York, 1990, pp. 703–718.
219. Riley, W.J. and Densford, J.W., Processes, Techniques, and Tools: The "How" of a Successful Design Control System, *Medical Device and Diagnostic Industry Magazine*, Vol. 19, No. 10,1997, pp. 75–80.
220. Riordan, J.J., Making Sure that Medical Devices Work, *Journal of the Association for the Advancement of Medical Instruments*, Vol. 6, No. 2, March 1972.

221. Ritzel, K., Industrial Design Moves Beyond Ergonomics, *Medical Device and Diagnostic Industry Magazine*, Vol. 19, No. 11, 1997, pp. 34–38.
222. Rose, H.B., A Small Instrument Manufacturer's Experience with Medical Instrument Reliability, *Proceedings of the Annual Reliability and Maintainability Symposium*, 1972, pp. 251–254.
223. Russell, C., Human Error: Avoidable Mistakes Kill 100,000 Patients a Year, *The Washington Post*, February 18, 1992.
224. Sahni, A., Seven Basic Tools that Can Improve Quality, *Medical Device and Diagnostic Industry Magazine*, Vol. 20, No. 4, 1998, pp. 89–96.
225. Saltos, R., Man Killed by Accident with Medical Radiation, *The Boston Globe*, June 20, 1986, p. 1.
226. Santel, C., Trautmann, C., and Liu, W.,The Integration of a Formal Safety Analysis into the Software Engineering Process: An Example from the Pacemaker Industry, *Proceedings of the Symposium on the Engineering of Computer Based-Medical Systems*, 1988, pp. 152–154.
227. Sawyer, D., An Introduction to Human Factors in Medical Devices, Center for Devices and Radiological Health, Food and Drug Administration, Washington, DC, 1996.
228. Schneider, P. and Hines, M.L.A., Classification of Medical Device Software, *Proceedings of the Symposium on Applied Computing*, 1990, pp. 20–27.
229. Schreiber, P., Human Factors Issues with Anesthesia Devices, *Proceedings of the First Symposium on Human Factors in Medical Devices*, 1989, pp. 32–36.
230. Schwartz, A.P., A Call for Real Added Value, *Medical Industry Executive*, February/March 1994, pp. 5–9.
231. Schwartz, R.J., Weiss, K.M., and Buchanan, A.V., Error Control in Medical Data, *MD Computing*, Vol. 2, No. 2, 1985, pp. 19–25.
232. Senders, J.W., Medical Devices, Medical Errors, and Medical Accidents, in *Human Error in Medicine*, edited by M.S. Bogner, Lawrence Erlbaum Associates, Publishers, Hillsdale, NJ, 1994, pp. 159–177.
233. Serig, D.I., Radiopharmaceutical Misadministrations: What's Wrong, in *Human Error in Medicine*, edited by M.S. Bogner, Lawrence Erlbaum Associates Publishers, Hillsdale, NJ, 1994, pp. 179–195.
234. Shaw, R., Safety Critical Software and Current Standards Initiatives, in *Computer Methods and Programs in Biomedicine*, Vol. 44, No. 1, July 1994.
235. Shephard, M.D., Shepherd's System for Medical Device Incident Investigation, Quest Publishing Company, Brea, CA, 1992.
236. Shepherd, M., A Systems Approach to Medical Device Safety, *Monograph*, Association for the Advancement of Medical Instrumentation, Arlington, VA, 1983.
237. Shepherd, M. and Brown, R., Utilizing a Systems Approach to Categorize Device–Related Failures and Define User and Operator Errors, *Biomedical Instrumentation and Technology*, November/December, 1992, pp. 461–475.
238. Sherertz, R.J. and Streed, S.A., Medical Devices: Significant Risk vs. Nonsignificant Risk, *JAMA*, Vol. 272, No. 12, 1994, pp. 955–956.

239. Sheridan, T.B. and Thompson, J.M., People Versus Computers in Medicine, in *Human Error in Medicine*, edited by M.S. Bogner, Lawrence Erlbaum Associates, Publishers, Hillsdale, NJ, 1994, pp. 141–158.
240. Simonaitis, D.F., Anderson, R.T., and Kaye, M.P., Reliability Evaluation of a Heart Assist System, *Proceedings of the Annual Reliability and Maintainability Symposium*, 1972, pp. 233–241.
241. Smirnov, I.P. and Shneps, M.A., Medical System Engineering, *Proceedings of the IEEE*, Vol. 57, No. 11, 1969, pp. 1869–1879.
242. Smith, C.E. and Peel, D., Safety Aspects of the Use of Microprocessors in Medical Equipment, *Measurement and Control*, Vol. 21, No. 9, 1988, pp. 275–276.
243. Stahlhurt, R.W., To Err is Human: Human Error in Medicine is Common, Inevitable and Manageable, *Medical Device and Diagnostic Industry Magazine*, Vol. 19, No. 11, 1997, pp. 13–14.
244. Stahlhut, R.W., Designing as if People Matte*red, American Medical Student Association Task Force Quarterly*, 1995, pp. 37–38.
245. Stanley, P.E., Monitors that Save Lives Can Also Kill, *Modern Hospital*, Vol. 108, No. 3, 1967, pp. 119–121.
246. Statistical Guidance for Clinical Trials of Non-Diagnostic Medical Devices, Center for Devices and Radiological Health, Food and Drug Administration, Rockville, MD, 1996.
247. Stein, P.E., Reliability and Performance Criteria for Electromedical Apparatus, *Proceedings of the Annual Reliability and Maintainability Symposium*, 1973, p. 89.
248. Steindel, B., Quality Control in the Practice of Medicine, *Proceedings of the 11th Annual West Coast Reliability Symposium*, 1970, pp. 197–202.
249. Stephens, R.N., *Medical Device Vigilance/Monitoring*, Interpharm Press, Buffalo Grove, IL, 1997.
250. Steward, S., Making Device Software Truly Trustworthy, *Medical Device and Diagnostic Industry Magazine*, Vol. 20, No. 1, 1998, pp. 86–89.
251. Subramanian, S., Elliott, L., Vishnuvajjala, R.V., Tsai, W.T., and Mojdejbakhsh, R., Fault Mitigation in Safety-Critical Software Systems, *Proceedings of the 9th IEEE Symposium on Computer-Based Medical Systems*, 1996, pp. 12–17.
252. Swartz, E.M., Product Liability, Manufacturer Responsibility for Defective or Neligently Designed Medical and Surgical Instruments, *De Paul Law Review*, Vol. 18, 1969, pp. 348–407.
253. Szycher, M., The Medical Device Industry, *Journal of Biomaterials Applications*, Vol. 11, 1996, pp. 76–118.
254. Taylor, E.F., Reliability, Risk and Reason in Medical Equipment, *Proceedings of the 5th Annual Meeting of the Association for the Advancement of Medical Instrumentation*, March 23, 1970, pp. 1–5.
255. Taylor, E.F., Reliability: What Happens if…, *Proceedings of the Annual Symposium on Reliability*, 1969, pp. 20–24.
256. Taylor, E.F., The Effect of Medical Test Instrument Reliability on Patient Risks, *Proceedings of the Annual Symposium on Reliability*, 1969, pp. 328–330.

257. Taylor, E.F., The Impact of FDA Regulations on Medical Devices, *Proceedings of the Annual Reliability and Maintainability Symposium*, 1980, pp. 8–10.
258. Taylor, E.F., The Reliability Engineer in the Health Care System, *Proceedings of the Reliability and Maintainability Symposium*, 1972, pp. 245–248.
259. Thibeault, A., Documenting a Failure Investigation, *Medical Device and Diagnostic Industry Magazine*, Vol. 19, No. 10, 1997, pp. 14,15.
260. Thibeault, A., Handling Reports of Product use in Incidents Causing Injury or Death, *Medical Device and Diagnostic Industry Magazine*, Vol. 20, No. 9, 1998, pp. 91–96.
261. Thompson, C.N.W., Model of Human Performance Reliability in Health Care System, *Proceedings of the Annual Reliability and Maintainability Symposium*, 1974, pp. 335–339.
262. Thompson, P.W., Safer Design of Anaesthesia Equipment, *Br. J. Anaesth.*, Vol. 59, 1987, pp. 913–921.
263. Thompson, R.C., Fault Therapy Machines Cause Radiation Overdoses, *FDA Consumer*, Vol. 21, No. 10, 1987, pp. 37–38.
264. Tracking Medical Errors, From Humans to Machines, *The New York Times*, March 31, 1992, p. 81.
265. Truby, C., Quality and Productivity: Then and Now, *Medical Device and Diagnostic Industry Magazine*, Vol. 21, No. 6, 1999, pp. 104–108.
266. Van Cott, H., Human Errors: Their Causes and Reduction, in *Human Error in Medicine*, edited by M.S. Bogner, Lawrence Erlbaum Associates Publishers, Hillsdale, NJ, 1994, pp. 53–91.
267. Voas, J., Miller, K., and Payne, J., A Software Analysis Technique for Quantifying Reliability in High Risk Medical Devices, *Proceedings of the 6th Annual IEEE Symposium on Computer-Based Medical Systems*, 1993, pp. 64–69.
268. Vroman, G., Cohen, I., and Volkman, N., Misinterpreting Cognitive Decline in the Elderly: Blaming the Patient, in *Human Error in Medicine*, edited by M.S. Bogner, Lawrence Erlbaum Associates Publishers, Hillsdale, NJ, 1994, pp. 93–122.
269. Wadsworth, H.M., Standards for Tools and Techniques, *Transactions of the ASQC Annual Conference*, 1994, pp. 882–887.
270. Waits, W., Planned Maintenance, *Med. Res. Eng.*, Vol. 7, No. 12, December 1968.
271. Walfish, S., Using Statistical Analysis in Device Testing, *Medical Device and Diagnostic Industry Magazine*, Vol. 19, No. 9, 1997, pp. 36–42.
272. Waynant, R.W., Quantitative Risk Analysis of Medical Devices: An Endoscopic Imaging Example, *Proceedings of the International Society for Optical Engineering Conference*, 1995, pp. 237–245.
273. Wear, J.O., Maintenance of Medical Equipment in the Veterans Administration, *Proceedings of the Third Annual Meeting of the Assoc. for the Advan. of Med. Inst.*, July 1968, pp. 10–14.

274. Weese, D.L. and Buffaloe, V.A., Conducting Process Validations with Confidence, *Medical Device and Diagnostic Industry Magazine*, Vol. 20, No. 1, 1998, pp. 107–112.
275. Weide, P., Improving Medical Device Safety with Automated Software Testing, *Medical Device and Diagnostic Industry Magazine*, Vol. 16, No. 8, 1994, pp. 66–79.
276. Weinberg, D.I., Artley, J.A., Whalen, R.E., and McIntosh, M.T., Electrical Shock Hazards in Cardiac Catheterization, *Circ. Research*, Vol. 11, 1962, pp. 1004–1011.
277. Weinger, M.B. and Englund, C.E., Ergonomic and Human Factors Affecting Anesthetic Vigilance and Monitoring Performance in the Operating Room Environment, *Anesthesiology*, Vol. 73, No. 5, 1990, pp. 995–1021.
278. Whalen, R.E., Starmer, C.F., and McIntosh, H.D., Electrical Hazards Associated with Cardiac Pacemaking, *Transactions of the New York Academy of Sciences*, Vol. 111, 1964, pp. 922–931.
279. Whelpton, D., Equipment Management: The Cinderella Bio-Engineering, *J. Biomed. Eng.*, Vol. 10, 1988, pp. 499–505.
280. Wiklund, M.E., Human Error Signals Opportunity for Design Improvement, *Medical Device and Diagnostic Industry Magazine*, Vol. 14, No. 2, 1992, pp. 57–61.
281. Wiklund, M.E., Making Medical Device Interfaces More User Friendly, *Medical Device and Diagnostic Industry Magazine*, Vol. 20, No. 5, 1998, pp. 177–183.
282. Wiklund, M.E., *Medical Device and Equipment Design*, Interpharm Press, Buffalo Grove, IL, 1995.
283. Wilcox, S.B., Building Human Factors Into the Design of Medical Products, *Proceedings of the First Symposium on Human Factors in Medical Devices*, 1989, pp. 14–16.
284. Willingmyre, G.T., Industry's Role in Standards Development, in *The Medical Device Industry*, edited by N.F. Estrin, Marcel Dekker Inc., New York, 1990, pp. 139–150.
285. Willis, G., Failure Modes and Effects Analysis in Clinical Engineering, *Journal of Clinical Engineering*, Vol. 17, 1992, pp. 59–62.
286. Wilson, C.J., ME: A Survey of Anesthetic Misadventures, *Anaesthesia*, Vol. 36, 1981, pp. 933–936.
287. Winger, J., Bray, T., and Halter, P., Lower Health Costs from High Reliability, *Proceedings of the Annual Reliability and Maintainability Symposium*, 1979, pp. 203–210.
288. Wood, B.J. and Ermes, J.W., Applying Hazard Analysis to Medical Devices, Part II, *Medical Device and Diagnostic Industry Magazine*, Vol. 15, No. 3, 1993, pp. 58–64.
289. Wood, B.J., Software Risk Management for Medical Devices, *Medical Device and Diagnostic Industry Magazine*, Vol. 21, No. 1, 1999, pp. 139–145.

290. Young, F.E., Validation of Medical Software: Present Policy of the Food and Drug Administration, *Annals of Internal Medicine*, Vol. 106. 1987, pp. 628–629.

291. Zane, M., Patient Care Appraisal, *Proceedings of the Annual Reliability and Maintainability Symposium*, 1976, pp. 84–91.

Index

S

T

U

V

W

X